# THE ALTENBERG 16:

# AN EXPOSÉ OF THE EVOLUTION INDUSTRY

# About the Author

Suzan Mazur's interest in evolution began with a flight from Nairobi into Olduvai Gorge to interview the late paleoanthropologist Mary Leakey. Because of ideological struggles, the Kenyan-Tanzanian border was closed, and Leakey was the only reason authorities in Dar es Salaam agreed to give landing clearance. The meeting followed discovery by Leakey and her team of the 3.6 million-year-old hominid footprints at Laetoli. Suzan Mazur's reports have since appeared in the *Financial Times, The Economist, Forbes, Newsday, Philadelphia Inquirer, Archaeology, Connoisseur, Omni* and others, as well as on *PBS, CBC* and *MBC*. She has been a guest on *McLaughlin, Charlie Rose* and various *Fox Television News* programs.

*Photo: Leonid Lubianitsky*

# THE ALTENBERG 16:

# AN EXPOSÉ OF THE EVOLUTION INDUSTRY

***SUZAN MAZUR***

North Atlantic Books
Berkeley, California

**SCOOP MEDIA**
Wellington, New Zealand

Published by
North Atlantic Books
P.O. Box 12327
Berkeley, California 94712

Printed in the United States of America.

*The Altenberg 16: An Exposé of the Evolution Industry* is sponsored by the Society for the Study of Native Arts and Sciences, a nonprofit educational corporation whose goals are to develop an educational and cross-cultural perspective linking various scientific, social, and artistic fields; to nurture a holistic view of arts, sciences, humanities, and healing; and to publish and distribute literature on the relationship of mind, body, and nature.

North Atlantic Books' publications are available through most bookstores. For further information, call 800-733-3000 or visit our Web site at www.northatlanticbooks.com.

Library of Congress Cataloging-in-Publication Data
Mazur, Suzan.
The Altenberg 16 : an expose of the evolution industry / Suzan Mazur.
p. cm.
"First published by Scoop Media, Wellington, New Zealand, 2009."
ISBN: 978-1-55643-924-7
1. Evolution (Biology) 2. Evolution (Biology)—Research—Social aspects.
I. Title.
QH360.5.M35 2009
576.8—dc22

2009042397

1 2 3 4 5 6 7 8 9 UNITED 16 15 14 13 12 11 10

# Contents

# Foreword

"Olduvai was first found in 1911 by a German naturalist called Professor Katwinkle. That was when Tanzania was German East Africa – a German colony.

Katwinkle took back some fossils to Berlin. And the Germans sent out an expedition in 1913. But then, of course, the First World War happened so they couldn't come back.

But my husband came out for the first time in 1930. He was here in 1930-31 with the German geologists who worked here. And then, of course, there was a hiatus for a long time.

I came here first in 1936, no, 1935. But in those days we had very little money and most we could do was come here for a week or two and work rather hurriedly. And it wasn't until the discovery of the *Zinjanthropus*... *Zinjanthropus* was in 1959.

We were coming from tents in Nairobi. That was before we had any substantial grants from the National Geographic Society. And we were prospecting, that is I was prospecting. My husband wasn't feeling well.

The site actually is just below this camp here. I saw a bit of bone that looked strange and it was in fact this mastoid skull. And when I brushed some soil away, I found some teeth which showed it was hominid straight away. So I fetched my husband down from the other camp. In those days we camped on the other side of the Gorge.

I was very excited because we'd been looking for it for a long time. And we were very fortunate because Des Bartlett, a professional photographer, was due to come down and photograph the work here.

So we covered it up and waited until he arrived two days later. So we have a film of the whole of the uncovering of the skull, which is very fortunate.

The discovery of the *Zinjanthropus* skull created a stir. And following that, the National Geographic Society decided to fund the work at Olduvai, which they still do to this day."

**Mary Leakey**, paleoanthropologist and media star

*Conversation with Suzan Mazur, Olduvai Gorge, Tanzania, June 1980*

# Introduction

*"There has never been a theory of evolution."*

– Cytogeneticist **Antonio Lima-de-Faria**,
*Evolution without Selection*

This book, *The Altenberg 16: An Exposé of the Evolution Industry*, looks at the rivalry in science today surrounding attempts to discover the elusive process of evolution, as rethinking evolution is pushed to the political front burner in hopes that "survival of the fittest" ideology can be replaced with a more humane explanation for our existence and stave off further wars, economic crises and destruction of the Earth.

Evolutionary science is as much about the posturing, salesmanship, stonewalling and bullying that goes on as it is about actual scientific theory. It is a social discourse involving hypotheses of staggering complexity with scientists, recipients of the biggest grants of any intellectuals, assuming the power of politicians while engaged in *Animal House* pie-throwing and name-calling: "ham-fisted", "looney Marxist hangover", "secular creationist", "philosopher" (a scientist who can't get grants anymore), "quack", "crackpot". . .

In short, it's a modern day quest for the holy grail, but with few knights. At a time that calls for scientific vision, scientific inquiry's been hijacked by an industry of greed, with evolution books hyped like snake oil at a carnival.

Perhaps the most egregious display of commercial dishonesty is this year's celebration of Charles Darwin's *On the Origin of Species* – the so-called theory of evolution by natural selection, *i.e.*, survival of the fittest, a brand foisted on us 150 years ago.

Scientists agree that natural selection can occur. But the scientific community also knows that natural selection has little to do with long-term changes in populations. And that self-assembly and self-organization are real, that is, matter can form without a genetic recipe.

The snowflake (non-living) assembles this way. And the hydra (living), for example, can re-assemble its scattered cells even after being forced through a sieve. Yet, many scientists term self-assembly and self-organization "woo woo".

Coinciding with the 2009 Darwinian celebration, MIT will publish a book by 16 biologists and philosophers who met at Altenberg, Austria's Konrad Lorenz Institute in July 2008 to discuss a reformulation of the theory of evolution. That's the mansion made famous by Konrad Lorenz's imprinting experiments, where he got his geese to follow him because they sensed he was their mother.

The symposium's title was "Toward an Extended Evolutionary Synthesis?", although the event has actually helped kick off an evolution remix.

Some of "the Altenberg 16" (as I've dubbed the group) have said they're trying to steer evolutionary science in a more honest direction, that is, by addressing non-centrality of the gene. They say that the Modern Synthesis, also called neo-Darwinism – which cobbled together the budding field of population genetics and paleontology, etc., 70 years ago – also marginalized the inquiry into morphology. And that it was then – in the 1930s and 1940s – that the seeds of corruption were planted and an "evolution industry" was born.

I broke the story about the Altenberg meeting in March 2008 with the assistance of Alastair Thompson and the team at *Scoop Media*, the independent news agency based in New Zealand. (Chapter 2, "Altenberg! The Woodstock of Evolution?") *Science* magazine later noted that my reporting "reverberated throughout the evolutionary biology community."

But will the Altenberg 16 deliver? Will they and their findings help rid us of the natural selection "survival of the fittest" mentality that has plagued civilization for a century and a half, and on which Darwinism and neo-Darwinism are based, now that the cat is out of the bag that natural selection is largely a political brand? That it is not THE process of evolution. That it is a tool left over from 19th century British imperial exploits.

Certain things look promising. First, while the Altenberg 16 have roots in neo-Darwinian theory, they recognize the need to challenge the prevailing Modern Synthesis because there's too much it doesn't explain.

For example, the Modern Synthesis was produced when genetics was still a baby and scientists have now discovered all the human genes there are to be found. We've only got 20,000 - 25,000 of them (some estimates range from 15,000 - 30,000). So there's a push for more investigation into non-genetic areas, for how body plans originated, for instance.

Second, the Extended Synthesis symposium was hosted by Konrad Lorenz Institute, where for years there have been discussions about self-organization.

Third, one of the stars of the conference, New York Medical College cell biologist Stuart Newman hypothesizes that all 35 or so animal phyla physically self-organized by the time of the Cambrian explosion (a half billion years ago) without a genetic program. Hardwiring, *i.e.*, selection supposedly followed.

Fourth, KLI's chairman, Gerd Müller has collaborated with Stuart Newman on a book about the origin of form. And Newman has other allies within the group, including Yale biologist Gunter Wagner, Budapest biologist and KLI board member Eors Szathmary, as well as KLI's science manager, Werner Callebaut – a Belgian philosopher who delivered the non-centrality of gene paper.

I published a "first peek" at Stuart Newman's concept (Appendix B, "Stuart Newman's High Tea") following his presentation at the University of Notre Dame in March 2008.

There has been a stonewalling on science blogs about self-organization (the umbrella term "plasticity" is now creeping in post-Altenberg to explain the self-organization process). The consensus of the evolution pack still seems to be that if an idea doesn't fit in with Darwinism and neo-Darwinism – *keep it out.*

Meanwhile, Swedish cytogeneticist Antonio Lima-de-Faria, author of the book *Evolution without Selection,* sees any continuance of the natural selection concept as "compromise". He says Darwinism and neo-Darwinism deal only with the biological or "terminal" phase of evolution and impede discovery of a real mechanism, which is "primaeval" – based on elementary particles, chemical elements and minerals (Chapter 8, "Knight of the North Star").

Lima-de-Faria's views are considered "extreme" by some in the scientific establishment even 20 years after the publication of *Evolution without Selection* about self-assembly – which he defines as "the spontaneous aggregation of biological structures involving formation of weak chemical bonds between surfaces with complementary shapes". However, it appears others may be warming up to concepts he laid down decades ago, as evidenced by comments at the 2008 World Science Festival in New York.

**Steve Benner**, pioneer of synthetic biology and founder, Westheimer Institute for Science:

> "But certainly our view of how life originated on Earth is very much dependent on minerals being involved in the process to control the chemistry. . . . So in that sense, I agree with my

> distinguished colleague from Lund [Lima-de-Faria]."

**Paul Davies**, theoretical physicist and astrobiologist, Director BEYOND Center, Arizona State University:

> "There has to be a pathway from chemistry to biology–powerful levels before Darwinian evolution even kicks in."

Lima-de-Faria notes that when Charles Darwin's *Origin of Species* and Alfred Russel Wallace's essay on natural selection came out, both were criticized. He quotes Darwin quoting a Professor Haughton of Dublin "that everything new in them was false and what was true was old".

Lima-de-Faria adds that "time and again, any radically new approach" in science is met with the same response.

The commercial media is both ignorant of and blocks coverage of stories about non-centrality of the gene because its science advertising dollars come from the gene-centered Darwin industry. With declining ad revenue already widespread, and employee layoffs and contract buyouts in the editorial departments of news organizations like *Newsweek, Time,* the *Washington Post* as well as the *New York Times* – reporting on an evolution paradigm shift could mean the loss of even more advertising and/or yet another editor's job.

But neither will most science blogs report that there's a paradigm shift afoot because they share the same ideology as the corporate media. At the same time, the Darwin industry is also in bed with government, even as political leaders remain clueless about evolution.

Thus, the public is unaware that its dollars are being squandered on funding of mediocre, middlebrow science or that its children are being intellectually starved as a result of outdated texts and unenlightened teachers.

However, while the Altenberg 16 organizers noted that their July symposium "could turn into a major stepping stone for the entire field of evolutionary biology," this book is not an endorsement of any attempt to "graft" novel ideas onto the Modern Synthesis – only of the decision to begin sorting out the mess. The real task is one of making a coherent theory of evolution, including pre-biotic evolution, where none previously existed. That will require casting a wide net for visionaries who have political courage.

# Chronology

**October 2007** – Publication of philosopher Jerry Fodor's *London Review of Books* story "Why Pigs Don't Have Wings" on the subject of evolution without adaptation.

**December 2007** – Evolutionary biologist Massimo Pigliucci publishes a story in the journal *Evolution* asking if we need an Extended Evolutionary Synthesis.

**February 2008** – Massimo Pigliucci leaks word of a July 2008 meeting about an evolution remix at Altenberg during a phone conversation with this author and again during a meeting in person in Manhattan.

**February 2008** – The Konrad Lorenz Institute in Altenberg leaks its letter of invitation to me about the remix event.

**March 4, 2008** – My story "Altenberg! The Woodstock of Evolution?" is syndicated by *Scoop Media,* the independent news agency based in New Zealand.

**March 2008** – Stuart Newman gives an origin of form talk during "high tea" at the University of Notre Dame, introducing his concept of DPMs (dynamical patterning modules) as the pattern language modern animals used to physically self-organize without a genetic program by the time of the Cambrian explosion.

**March 2008** – Author Richard Dawkins at Tribeca Barnes & Noble addresses the subject of Altenberg and a reformulation of the theory of evolution.

**April 2008** – Richard Dawkins announces during a Q&A at a New York Ethical Culture talk that he is renouncing Darwinism as religion.

**April 2008** – Stuart Newman and co-author Ramray Bhat publish their theory of form paper in *Physical Biology* making it freely available to the public.

**May 2008** – Rockefeller University Evolution symposium.

**May 2008** – Massimo Pigliucci – The Secular Humanist Society of New York – paradigm shift talk.

**May/June 2008** – World Science Festival – New York.

**July 10-13, 2008** – Konrad Lorenz Institute – Extended Synthesis Symposium at Altenberg.

**July 11, 2008** – *Science* magazine picks up the Altenberg story, announcing that my March 4 article has "reverberated throughout the evolutionary biology community".

**September 18, 2008** – *Nature* magazine leads with my Altenberg-Woodstock theme (no attribution) in its cover story, running a correction with credit to me on October 2.

**March 3-7, 2009** – Rome evolution conference: "Biological Evolution: Facts and Theories – A Critical Appraisal 150 Years After *The Origin of Species*" sponsored by the Pontifical Council for Culture. "Evolutionary Mechanisms" are the focus of a full day of talks.

# Evolution Tribes

(partial list)

**FORM & STRUCTURE**

**Ricardo Azevedo** – complexity investigator, University of Houston, US

**Lev Beloussov** – chair of embryology department, Moscow State University, Russia

**Ramray Bhat** – complexity theorist, New York Medical College, US

**Mary Jane West-Eberhard** – biologist, University of Costa Rica

**Gabor Forgacs** – University of Missouri biological physicist, US

**Scott Gilbert** – biologist, Swarthmore College, US

**Brian Goodwin** (1931 – 2009) – theoretical biologist, Schumacher College, UK

**Pegio-Yukio Gunji** – theorist, Kobe University, Japan

**Leendert Van Der Hammen** – structuralist, The Netherlands

**Mae-Wan Ho** - geneticist and director, Institute of Science in Society (ISIS), UK

**Kiyohiko Ikeda** – Waseda University, Japan

**Stuart Kauffman** – complexity pioneer and philosopher, University of Calgary, Canada

**Dave Lambert** – structuralist, Massey University, New Zealand

**Antonio Lima-de-Faria** – emeritus professor of molecular cytogenetics, Lund University, Sweden

**Mark McMenamin** – geologist, Mount Holyoke College, US

**Gerd Müller (A-16)** – chair of Konrad Lorenz Institute, Altenberg, Austria and chair of theoretical biology department, University of Vienna, Austria

**Stuart Newman (A-16)** – professor of cell biology and anatomy, New York Medical College, US

**Stuart Pivar** – torus investigator, New York

**Isaac Salazar-Ciudad** – morphological mechanism of evolution investigator, University of Helsinki, Finland

**Giuseppe Sermonti** – professor emeritus of genetics, University of Perugia, Italy

**Rupert Sheldrake** – biochemist, morphic fields and morphic resonance investigator, UK

**Atuhiro Sibatani** – molecular biologist, Japan

**Robert Ulanowicz** – professor emeritus, theoretical ecology, University of Maryland, US

**Vladimir Voeikov** – professor bioorganic chemistry, Moscow State University, Russia

**Gunter Wagner (A-16)** – Yale University evolutionary biologist, US

**Gerry Webster** – structuralist, University of Sussex, UK

## BIOLOGY & GENETICS

**Steve Benner** – pioneer of synthetic biology and founder, Westheimer Institute for Science

**Sean Carroll** – biologist, University of Wisconsin-Madison, US

**Jerry Coyne** – evolutionary geneticist, University of Chicago, US

**Jeffrey Feder** – geneticist, University of Notre Dame, US

**Doug Futuyma** – evolutionary biologist, SUNY – Stony Brook, US

**Sergey Gavrilets (A-16)** – biomathematician, University of Tennessee, US

**Eva Jablonka (A-16)** – geneticist (epigenetics), Tel Aviv University, Israel

**David Jablonski (A-16)** – professor of geophysical science, University of Chicago, US

**Marc Kirschner (A-16)** – professor of systems biology, Harvard Medical School, US

**Richard Lewontin** – Harvard University geneticist, US

**Michael Lynch** – evolutionary biologist, University of Indiana, US

**Lynn Margulis** – geoscientist (symbiosis), University of Massachusetts-Amherst, US and Oxford University, UK

**Larry Moran** – *Sandwalk* blogger and biochemist, University of Toronto

**Simon Conway Morris** – paleobiologist, Cambridge University, UK

**PZ Myers** – *Pharyngula* blogger and biologist

**F. John Odling-Smee (A-16)** – bioanthropologist, Oxford University, UK

**Michael Purugganan (A-16)** – evolutionary geneticist, New York University, US

**Eors Szathmary (A-16)** – biologist, Eotvos University and Collegium Budapest, Hungary

**Ulrich Technau** – evolution complexity investigator, University of Vienna, Austria

**David Sloan Wilson (A-16)** – evolutionary biologist, Binghamton University, US

**Greg Wray (A-16)** – biologist, Duke University, US

## ASTROBIOLOGY

**Roger Buick** – geologist, University of Washington, US

**David Deamer** – artificial life scientist, University of California–Santa Cruz, US

**Robert Hazen** – professor of earth sciences, George Mason University and NASA astrobiology team, Carnegie Institution of Washington, US

**Chris McKay** – planetary scientist, NASA Ames Research Center, US

**Bruce Runnegar** – paleontologist, University of California–Los Angeles, US

## PHILOSOPHY & PUBLIC INTELLECTUALS

**John Beatty (A-16)** – philosopher of biology, University of British Columbia, Canada

**Werner Callebaut (A-16)** – scientific manager of KLI, philosopher, Limburg University, Belgium

**Noam Chomsky** – linguist, Massachusetts Institute of Technology, US

**Richard Dawkins** – author and public intellectual, UK

**Niles Eldredge** – paleontologist, Museum of Natural History, US

**Jerry Fodor** – philosopher, Rutgers University, US

**Jean Gayon** – philosopher and historian, Universite Paris 1, France

**Donna Haraway** – philosopher, University of California – Santa Cruz, US

**David Koch** – philanthropist and principal private funder of PBS *Nova* series, Koch Industries, Inc., US

**Richard Leakey** – paleoanthropologist and author, SUNY Stony Brook, US

**Alan Love (A-16)** – philosopher, University of Minnesota, US

**Paul Nurse** – president, Rockefeller University, US

**Massimo Piattelli-Palmarini** – cognitive scientist, University of Arizona, US

**Massimo Pigliucci (A-16)** – philosopher and geneticist, SUNY Stony Brook, US

**Stan Salthe** – natural philosopher, Binghamton University, US

**Elliot Sober** – philosopher, University of Wisconsin-Madison, US

**"The Altenberg 16" (left to right):**

Sergey Gavrilets, Stuart Newman, David Sloan Wilson, John Beatty, Francis John Odling-Smee, Michael Purugganan, Greg Wray, David Jablonski, Marc Kirschner, Eors Szathmary, Gunter Wagner, Werner Callebaut, Eva Jablonka, Gerd Müller, Massimo Pigliucci, Alan Love. (Photo used with permission – use of this photo does not constitute an endorsement of this book by any of those pictured, nor by the Konrad Lorenz Institute.)

**Ramray Bhat** – New York Medical College

**Dave Deamer** – University of California – Santa Cruz

**Roger Buick** – University of Washington

**Niles Eldredge** – American Museum of Natural History

**Jerry Fodor** – Rutgers University

**Robert Hazen** – Carnegie Institution and George Mason University

**Scott Gilbert** – Swarthmore College

**Stuart Kauffman** – University of Calgary

**David H. Koch** – Science Philanthropist. *(Photo courtesy of David Koch)*

**Antonio Lima-de-Faria** – University of Lund

**Richard Lewontin** – Harvard University

**Chris McKay** – NASA's Ames Research Center

**Lynn Margulis** – Oxford University and the University of Massachusetts-Amherst
*(Photo by Jerry Bauer)*

**Stuart Newman** – New York Medical College
*(Photo by Jura Newman)*

**Gerd Müller** – University of Vienna and Chairman, Konrad Lorenz Institute

**Paul Nurse** – President, Rockefeller University

**Massimo Piattelli-Palmarini** – University of Arizona

**Bruce Runnegar** – University of California, Los Angeles

**Stuart Pivar** – New York

**Stan Salthe** – University of Binghamton

# Chapter 1

# The Altenberg 16

I had a phone conversation with SUNY Stony Brook biologist Massimo Pigliucci in February 2008. I was developing a piece for one of America's top ten newspapers about the evolution debate, which after many months of serious though informal discussion got killed. However, by the time the piece was stopped, I had already posted a provocative "rethinking evolution" series with *Scoop Media,* the "fiercely independent news agency" based in New Zealand run by Alastair Thompson, Selwyn Manning, and the Wellington team, David McLellan *et al.*

Pigliucci mentioned to me that there was to be a meeting of 16 scientists in Vienna in July to discuss the evolution debate and an Extended Evolutionary Synthesis. I thought little more about it at the time, since scientists meet regularly to talk about evolution. That is, until I actually met with Pigliucci on February 21 at the Pret A Manger cafeteria across from New York's 42nd Street library where he was doing research.

Massimo Pigliucci is a high-strung Italian-American with three PhDs – in botany, genetics and philosophy. He's 46 years old, with little trace of an Italian accent, having lived in the US for almost 20 years.

Pigliucci was dressed down in a windbreaker and jeans. He's got a wiry body and walks as rapidly as he speaks. There's a hint of red in his brush-like receding hair. Pigliucci was carrying a birthday gift and smiling.

He answered my questions and also gave me a bit of a lecture about self-organization and self-assembly (matter can form without a genetic recipe), by referring to the water I was drinking.

"What you're drinking now – water. Now we know that water is made of two atoms of hydrogen and one of oxygen. And yet the physical and chemical properties of water – the temperature at which it's going to freeze, the density that it has at certain temperatures and so on – cannot be derived directly from the knowledge of the physical and chemical properties of hydrogen and oxygen. In other words, water is an emergent property of a specific combination of hydrogen and oxygen.

The emergent properties are due to the fact that these atoms are not just mixed together in a two to one proportion. They're actually connected in molecules that have a particular shape. And it is that shape that is determined by the chemical bonds that in turn determine the physical and chemical properties. Emergence is a big component of self-organization. It's fundamental. It's very important. We don't even understand things like water."

Pigliucci also confirmed that humans have about 25,000 genes, that we're not going to find any more, and that was a principal reason why self-organization was currently an important focus of research.

He told me his investigations into epigenetics – the study of heritable changes in gene expression and function outside genetic variation – suggested there was a whole layer of inheritance on top of the genes which the current Modern Synthesis doesn't account for.

About ten days later, Pigliucci emailed me the paper he co-authored on "ecological epigenetics". The paper described how environmental change might affect fitness – previously an idea espoused by the controversial 19th century scientist Jean-Baptiste Lamarck.

One reviewer from the Faculty of 1000 Biology, a group that evaluates new papers in the field, commented on the Pigliucci *et al.* paper: "This perspective paper suggests that epigenetics may challenge the assumptions of Modern Evolutionary Synthesis."

Later that day Pigliucci sent a second email:

> "I just found out (see below) that the paper I sent you this morning has been nominated by the Faculty of 1000 as one of the most interesting recent papers in biology. Not a bad accomplishment, but more importantly a sure sign that people are beginning to pay attention to these things."

However, I'd also had a phone conversation with esteemed Harvard geneticist Richard Lewontin and referred to the conversation about epigenetics with Pigliucci. Lewontin had reservations about the vagueness of language.

Quoting Pigliucci, I told Lewontin one argument was that "we're beginning to have mounting empirical evidence to suspect that there's a whole additional layer of inheritance. Not just the genes."

To which Lewontin responded: "Suspect or know?"

Lewontin is not supportive of an Extended Synthesis, asking, "Why would we want to do that?" In fact, he thinks it's all a race to become the next Charles Darwin. (Chapter 7, "The One and Only Richard Lewontin")

Nevertheless, the Pret A Manger chat with Massimo Pigliucci was pleasant, even though Pigliucci was determined to stay on point. He had no interest in the antiquities trial in Rome, for instance, which I'd been following. And he got red-faced when I brought up Stuart Pivar, the independent scientist and old friend of Andy Warhol who has a theory that the bagel-shaped torus is the origin of form.

Pivar, knowing that Andy couldn't draw yet became the most famous pop artist of his time, has long been suspicious of credentialed evolutionary scientists with hefty foundation research grants.

Pigliucci again brought up the subject of the Austrian talks at our meeting and suggested I speak with organizers in Europe to see if press were invited. Later realizing something extraordinary might be brewing, I contacted the Konrad Lorenz Institute in Altenberg.

I got Werner Callebaut on the phone. Callebaut is a Belgian philosopher and KLI's scientific manager. He was friendly – like his autobiography on the institute's web site – and told me that he knew the paper I was writing the evolution debate piece for. He also said that one or two journalists did attend KLI sessions sometimes.

I next received the letter of invitation from KLI originally sent to the 16 scientists who I named "the Altenberg 16". It was signed by Massimo Pigliucci and KLI chairman Gerd Müller and described the talks as "a major event" and "a major stepping stone for the entire field of evolutionary biology".

On March 4, 2008, I published my first story about the Altenberg meeting to reformulate the Modern Synthesis or neo-Darwinian theory, mentioning there would be an MIT book as well. The story created a sensation. Because of the huge response, Altenberg organizers later said they would bar both media and the public from the event.

But by then I was familiar with the general content of many of the papers. I'd also written stories about perhaps Altenberg's most pivotal participant – Stuart Newman – who presented his hypothesis on the origin of form by self-organization (Chapter 4, "Theory of Form to Center Stage", Chapter 5, "The

Two Stus", also Appendix B). I'd been in touch with other Altenberg 16 scientists by phone and email. And I'd become increasingly curious about the evolution industry and so pursued the story...

The question remains, will Altenberg prove a "major stepping stone for the entire field" of evolutionary science?

Let's peek at what the Altenberg 16 were slated to present at KLI.

## SELECTION AND ADAPTATION REFORMED

### John Beatty – "chance variation"

John Beatty is a Texan with a PhD in the History and Philosophy of Science. As a philosopher of biology at the University of British Columbia, he has been investigating the connection between biology and the State, as in the Manhattan Project. Beatty told me this subject would not get covered at Altenberg, however.

Hopefully Beatty's equally provocative research into "how evolutionary biology is as much like history as it is like science" did. (Is there a "chance" Beatty is referring to invention of the natural selection brand – "survival of the fittest" – by Darwin *et al.*?)

Beatty was originally going to discuss "drift", that is, gene frequencies over generations. He's said it's difficult or impossible to distinguish drift from natural selection.

In a phone conversation a couple of weeks before the Altenberg meeting, Beatty told me he'd been reassigned the topic of chance variation by Massimo Pigliucci. Beatty has co-authored the book *The Empire of Chance: How Probability Changed Science and Everyday Life*.

Here's the Beatty Altenberg abstract:

> "A topic of considerable interest and controversy concerns the contingency of evolutionary outcomes. Stephen Gould famously (or infamously) argued that replays of the "tape" of life will lead to widely different outcomes. And since then there have been many attempts to confirm or refute his thesis on the basis of controlled laboratory experiments, "natural experiments," and computer simulations. My concern is not so much to adjudicate the controversy, as to analyze it. I will focus on one of the two senses of "contingency" employed by Gould and others, namely, the unpredictability of evolutionary outcomes. And I will emphasize one particular source of unpredictability, namely random variation, or more specifically still, the historical order, in which random mutations occur. The contingency of evolutionary outcomes has been thought to undermine the "importance" of natural selection. One of the general questions at issue in my paper will be the meaning of claims about the importance of this or that evolutionary agent, and in particular, what is at stake in debates about the relative importance of selection vs. variation."

**Sergey Gavrilets – fitness landscapes (reproductive success visualized in terms of a range of mountains – peaks, valleys, flats – also called "adaptive landscapes")**

I reached Sergey Gavrilets by phone a couple of weeks before the Altenberg symposium. He made a point of telling me in dramatic Russian-English that he didn't like my March 4 Altenberg story, neither the tone nor substance, and objected to being compared to a Woodstock rock star. He said Altenberg was all about "normal developments" in biology.

Sergey Gavrilets has a PhD in physics and mathematics from Moscow State University, Russia and is a 2008 Guggenheim Fellow. He specializes in biomathematics at the University of Tennessee.

Gavrilets was originally scheduled to speak about neutralism, relating to species inhabiting the same environment, but not intentionally interacting, as in the tarantula and cactus – or maybe neo-Darwinist and self-organizationalist.

He said he was reassigned the topic of "fitness landscapes" by Massimo Pigliucci. Gavrilets is the author of the book *Fitness Landscapes and the Origin of Species,* in which he presents a "quantitative theory" for the origin of species using "simple math models".

Commenting on Gavrilets' book, Doug Futuyma, the celebrated evolutionary biologist from SUNY Stony Brook, wrote the following: "A deep reading of his book will require considerably more mathematical competence than most evolutionary biologists (including this reviewer) have. . . ."

Gavrilets was to also address "future" fitness landscapes.

Here's the Gavrilets abstract: "High-dimensional Fitness Landscapes and the Origins of Biodiversity":

> "The Modern Evolutionary Synthesis of the 1930s and 1940s became possible only after the development of theoretical population genetics by Fisher, Wright, and Haldane who built a series of mathematical models, approaches, and techniques demonstrating how natural selection, mutation, drift, migration, and other evolutionary factors shape genetic and phenotypic characteristics of biological populations.

One of the theoretical constructions that emerged at that time and has since proved extremely useful both in evolutionary biology and well outside of it is that of "fitness landscapes", which are also known as "adaptive landscapes", "adaptive topographies", and "surfaces of selective value". I will describe recent advances in the theory of fitness landscapes that explicitly account for the fact that biologically realistic fitness landscapes have extremely high dimensionality. I will also consider the implications of the theoretical results on the properties of high-dimensional landscapes for understanding the processes of the origin of species.

At the end I will discuss two research areas which are of great interest and importance both for scientists and general public. The first area has to do with the dynamics of adaptive radiation and large-scale evolutionary diversification. The second area concerns the ultimate speciation event – the origin of our own species. The progress in both these areas will likely depend on the success in building adequate mathematical theories."

**David Sloan Wilson – multilevel selection ("altruism and pro-social behaviors" evolve because they benefit the group, although they may disadvantage individuals in the group)**

Wilson is a Binghamton University evolutionary biologist and *Huffington Post* blogger.

He is the son of Sloan Wilson, author of the 1955 novel about the organization man, *The Man in the Grey Flannel Suit* – later portrayed on screen by Gregory Peck. DS Wilson is the author of *How Darwin's Theory Can Change the Way We Think about Our*

*Lives*. Wilson may be the one Altenberg 16 scientist most resistant to a coup d'etat re Darwinian natural selection. (Chapter 3, "Jerry Fodor and Stan Salthe Open the Evo Box").

## NEW VIEWS ON GENOMES AND INHERITANCE

### Greg Wray – gene regulatory networks (switching on and off of genes)

Director, Duke University Institute for Genome Sciences and Policy. Wray is a PhD in biology, and a bodybuilder who enjoys flexing his muscles for the camera.

Wray's Altenberg paper looks at altered interactions in gene regulatory networks and evolutionary transformations. He has said: "During embryonic development, a fertilized egg containing a few simple spatial cues is transformed into an intricately patterned, functioning organism consisting of thousands of differentiated cells." Wray is co-founder and associate editor of the journal *Evolution and Development*.

### Michael Purugganan – genomes and post-genomes

Director of New York University's Plant Evolutionary Genetics lab (2008 funding from the US Defense Department, National Science Foundation and Guggenheim Foundation). Purugganan, has a PhD in botany from the University of Georgia. His web site notes that he likes to dress in Armani, black Armani.

In March 2008, he emailed me that he was supplying the Altenberg 16 buttons for the talks at KLI.

The lab at NYU's Washington Square must rock. I ran into Xianfa Xie, a postdoc fellow from Purugganan's lab at the Rockefeller University Evolution event in May. As I approached the stage to have a word with the night's principal speaker, Xie was already grilling him on the subject of epigenetics. The speaker was Jerry Coyne, the University of Chicago geneticist and Richard Dawkins (*Selfish Gene* book) pal.

**Eva Jablonka – epigenetic inheritance (heritable changes in gene expression and function outside genetic variation)**

Geneticist, Tel-Aviv University. Eva Jablonka was the one woman presenting a paper at Altenberg. She does not appear to be intimidated by the flexing of muscles. Jablonka has been attacked in print as a "Marxist biologist" for her opposition to the Iraq War and ethnic cleansing and is quoted in *Israel-Academia-Monitor.com* defending her position:

> "I am sick and tired of hearing that the critical and concrete decision of the AUT and similar decisions are anti-Semitic. Like many others in Israel who have a rather close connection with the holocaust [Jablonka was born in Poland], I do not need to be preached at about this issue. I understand very well the sensitivity and paranoia of a people who like me, had families who perished and parents who still suffer the scars of the past, but I am also very worried about the misuses of our tragic history."

Eva Jablonka is the co-author of three books: *Evolution in Four Dimensions, Animal Traditions: Behavioral Inheritance in Evolution,* and *Epigenetic Inheritance and Evolution: The Lamarckian Dimension.*

**Francis John Odling-Smee – niche inheritance (organisms "inherit genes, and biotically transformed selection pressures in their environment from their ancestors")**

Oxford University lecturer in Bioanthropology. Odling-Smee's paper covers ecological inheritance in evolution. He is co-author of the book *Niche Construction: The Neglected Process in Evolution.*

## UNDERSTANDING THE PHENOTYPE

### David Jablonski – dynamics of macroevolution (mass-extinctions "set the stage" for "evolutionary recoveries")

Jablonski is said to be a very lively lecturer. He's a professor of geophysical science at the University of Chicago with a PhD from Yale and he chairs the Committee on Evolutionary Biology. Since the Altenberg meeting is "private", we will have to wait for the DVD sales to hear Jablonski speak about mass-extinctions and macroevoluion.

Jablonski *et al.* have established that multicellular animals had five major and a number of minor mass extinctions in a 530 million year span which "set the stage" for "evolutionary recoveries".

### Massimo Pigliucci – phenotypic plasticity (characteristics of an organism change in response to changes in environment)

Massimo Pigliucci was one of the coordinators of the Altenberg summit. He has three PhDs – in botany, genetics and philosophy. Pigliucci directs an ecology lab at the State University of New York, Stony Brook.

One Pigliucci associate has referred to Pigliucci's "flamboyance".

Pigliucci's web site "Rationally Speaking" carries the words of the Enlightenment's Marquis de Condorcet describing a public intellectual as one who devotes "him or herself to the tracking down of prejudices in the hiding places where priests, the schools, the government and all long-established institutions had gathered and protected them". But Pigliucci often prefers mudslinging to enlightenment.

I asked Pigliucci whether natural selection was politics or science during his May 2008 Secular Humanist Society of New York talk on "Paradigm Shifts and the Objectivity of Science."

He responded: **"Both"**.

## Stuart Newman – origin of form and body plans

Stuart Newman says Darwinism is a dying theory; it begins with life and doesn't address where form comes from. Newman thinks he has a coherent hypothesis for the origin of form. His paper served as the centerpiece of the Altenberg symposium.

Newman is a perfectionist right down to the last umlaut. He draws his inspiration from classical art. And he prefers penning commentaries on science and culture from upstate New York to the science wars of Manhattan.

Stuart Newman is Professor of Cell Biology and Anatomy at New York Medical College in Valhalla, where he also directs a lab. He's a KLI Fellow and external faculty member.

Newman's PhD is in chemical physics from the University of Chicago where he did post doctoral studies in theoretical biology, as well as at the School of Biological Sciences, University of Sussex.

His AB is from Columbia University. As a teen he studied science in Columbia's honors program taught by Richard Lewontin (see Chapter 7 – The One and Only Richard Lewontin).

Newman proposes that all 35 animal phyla self-organized by the time of the Cambrian explosion half a billion years ago using dynamical patterning modules (DPMs), a pattern language that called up certain physical processes and enabled highly plastic single-celled organisms to leap into multicellularity and build cavities, layers of tissue, segments, extremities, primitive hearts and even eyes.

Newman thinks selection followed as a "stabilizer" of form. (Chapter 5, "The Two Stus" and Appendix B)

He has collaborated with KLI Chair Gerd Müller and University of Missouri biological physicist Gabor Forgacs on aspects of his DPM hypothesis. Ramray Bhat is Newman's co-author on the DPM paper.

Newman has also co-authored the textbook *Biological Physics of the Developing Embryo* with Gabor Forgacs, and with Gerd Müller co-edited *Origination of Organismal Form: Beyond the Gene in Developmental and Evolutionary Biology*, a volume about the origination of body form during Ediacaran and early Cambrian periods.

## CONTRIBUTIONS FROM EVO-DEVO

### Gerd Müller – innovation (novel trait emerges and becomes fixed in population)

Chairman, Konrad Lorenz Institute for Evolution and Cognition Research, MD, PhD, Professor of Zoology and Chair, Department of Theoretical Biology at the University of Vienna where he is Director, Müller Lab.

Müller is also a referee for the Institute of Advanced Study, the Guggenheim Foundation, MacArthur Foundation and the NSF. With Stuart Newman, Müller co-edited the book *Origination of Organismal Form* (cited above) as well as "Special Issue: Evolutionary Innovation and Morphological Novelty" in the *Journal of Exp. Zoology* (2005). Newman cites Müller's collaboration in his DPM paper.

At the end of May, Gerd Müller sent the following polite email to me making it clear media were not invited to the symposium:

> *"Dear Suzan Mazur,*
>
> Thanks for your enthusiastic interest in the evolution debate and the workshop Massimo Pigliucci and I are organizing at the KLI this summer. I would like to point out that the KLI workshops, like many specialized meetings in the scientific domain, are based on personal invitation by the organizers. There is nothing secret or closed about this format, many meetings all over the world follow this procedure every day. In contrast with huge congresses, to which many hundreds of scientists flock to give a short lecture, the workshop format, where a handful of experts come together for several days, serves a very different purpose: These are working sessions that are meant to produce new results from the interaction among the participants. For this procedure to be productive, a small number of participants and an undisturbed setting are required. Even if we wanted to, we cannot accept additional attendances, neither by scientists nor by journalists, simply because the KLI is placed in a family mansion, and the room in which the workshops are held doesn't hold more than maybe 20 people. Given that the scientists working at the Institute at the time of a workshop have a right to listen in to the discussions, the room is already more than full. However, the organizers of the meeting, Massimo Pigliucci and

myself, as well as the participants, will be available for interviews after the meeting, when organizers can report about the actual event and its results.

Thanks, again, for your interest in this subject!

Sincerely,

Gerd Müller

Chairman KLI"

**Gunter Wagner – modularity ("Organisms seem to be organized into structural modules or "parts," presumably on account of a connection between parts and the ability to perform functions." – Dan McShea, Duke University)**

Gunter Wagner is Alison Richard Professor of Ecology and Evolutionary Biology, Yale University and a 1992 MacArthur Fellow. He's a native of Vienna with a PhD in Zoology from the University of Vienna and with postdoc research at Max-Planck Institute and the University of Gottingen.

Wagner has been at Yale since 1991 where he now directs a lab that "uses mathematical modeling to understand complex adaptations of organisms".

He's written a book called: *Modularity in Development and Evolution.*

Wagner did not respond to my attempts to reach him for an interview for my original Altenberg story. After the story appeared he emailed me as part of an Altenberg 16 group message (probably in error) saying, "Hey that is great! This is one button I will keep!" – referring to the Altenberg 16 buttons Michael Purruganan said he was making for the symposium.

**Marc Kirschner – evolvability (the body has a certain plasticity and can work around "errors")**

Professor of systems biology at Harvard Medical School and founding chair of the Department of Systems Biology. PhD University of California – Berkeley.

Marc Kirschner is recipient of the 2003 E.B. Wilson Medal, the American Society of Cell Biology's highest honor and is also a past president of ASCB. He's co-author of *The Plausibility of Life: Resolving Darwin's Dilemma*. And he's served as an adviser to the director of the National Institute of Health.

Kirschner's lab at HMS is investigating the mechanisms of establishing the basic vertebrate body plan and other areas.

Kirschner's Altenberg paper was on evolvability, meaning the body has a certain plasticity, where blood vessels, nerves, ligaments, etc. work around a so-called mutation.

## CHARACTERISTICS OF EXTENDED SYNTHESIS

**Werner Callebaut – non-centrality of the gene**

Werner Callebaut was the first to brief me on the particulars of the Altenberg meeting. Callebaut is scientific manager of KLI He has a PhD from the University of Ghent, Belgium and is a professor of philosophy, Limburg University, Belgium. Callebaut is the author of *Taking the Natural Turn, or How Real Philosophy is Done* and edited *The Vienna Series in Theoretical Biology* with Gerd Müller and Gunter Wagner.

Of all the Altenberg bios, Callebaut's is the most charming, with this glimpse into his 1950s primary school years growing up in a village in Flemish Brabant posted on the KLI web site:

"Most importantly, the school yard had one wall that was very high, and hence perfectly suited to show my pals who could pee higher."

Callebaut has been involved with the radio broadcasts of scientific conferences and was supportive of media attending Altenberg.

**Eors Szathmary – principles of transition**

A professor of biology at the Department of Plant Taxonomy and Ecology, Eotvos University and Collegium Budapest (Institute for Advanced Study), and a KLI board member. Szathmary co-authored two books with John Maynard Smith: *The Major Transitions in Evolution* and *The Origins of Life*.

Binghamton University philosopher Stan Salthe sent an email to me saying that "most folks would opt for John Maynard Smith for the accolade" of "most important evolutionary biologist of the passing generation", but that Harvard geneticist Richard Lewontin gets his vote. He thinks "Lewontin outclassed him [Smith] on originality and style." (Chapter 7, "The One and Only Richard Lewontin")

Szathmary served as president of the International Organisation for Systematic and Evolutionary Biology (1996-2002). Some of his "achievements" include: a mathematical description of phases of early evolution; a framework for discussing major transitions in evolution; a scenario for origin of genetic code.

Szathmary established the New Europe School for Theoretical Biology foundation to help Hungarian scientists find educational financing.

**Alan Love – The structure of evolutionary theory and biological knowledge**

Philosopher, University of Minnesota. PhD from the University of Pittsburgh in History and Philosophy of Science. Love's Altenberg paper was based on a presentation he gave in March 2008 at SUNY, Stony Brook – which was hosted by Massimo Pigliucci. Here is the Love abstract:

> "Much of the discussion about the adequacy of contemporary evolutionary theory has focused on its content, such as whether it integrates developmental considerations. A different approach is to explore the form or structure of evolutionary theory, which is in part a philosophical question about the nature of scientific theories. In this paper I adopt the latter route in order to introduce some epistemic materials for a 21st century synthesis. Specifically, I distinguish narrow and broad interpretations of evolutionary theory and argue that a broad interpretation is more appropriate for conceptualizing an expanded evolutionary synthesis (*e.g.*, one that includes development). This requires construing the structure of evolutionary theory as multiple problem domains exhibiting complex but coordinating relationships. As a consequence, we can observe a new perspective on the structure of biological knowledge and gain a concrete understanding of how 'nothing makes sense except in the light of evolution'."

# Chapter 2

# Altenberg! The Woodstock of Evolution?

*March 4, 2008*
*1:49 pm NZ*

It's not Yasgur's Farm, but what happens at the Konrad Lorenz Institute in Altenberg, Austria this July promises to be far more transforming for the world than Woodstock. What it amounts to is a gathering of 16 biologists and philosophers of rock star stature – **let's call them "the Altenberg 16"** – who recognize that the theory of evolution which most practicing biologists accept and which is taught in classrooms today, is inadequate in explaining our existence. It's pre the discovery of DNA, lacks a theory for body form and does not accomodate "other" new phenomena. So the theory Charles Darwin gave us, which was dusted off and repackaged 70 years ago, seems about to be reborn as the "Extended Evolutionary Synthesis".

Papers are in. MIT will publish the findings in 2009 – the 150th anniversary of Darwin's publication of *The Origin of Species*. And despite the fact that organizers are downplaying the Altenberg meeting as a discussion about whether there should be a new theory, it already appears a done deal. Some kind of shift away from the population genetic-centered view of evolution is afoot.

Indeed, history may one day view today's "Altenberg 16" as 19th century England's X Club of 9 – Thomas Huxley, Herbert Spencer, John Tyndall, *et al.* – who so shaped the science of their day.

Here then are **the Altenberg 16: John Beatty**, University of British Columbia; **Sergey Gavrilets**, University of Tennessee; **David Sloan Wilson**, Binghamton University; **Greg Wray**, Duke University; **Michael Purugganan**, New York University; **Eva Jablonka**, Tel-Aviv University; **John Odling-Smee**, Oxford University; **David Jablonski**, University of Chicago; **Massimo Pigliucci**, SUNY Stony Brook; **Stuart Newman**, New York Medical College; **Gerd Müller**, University of Vienna; **Gunter Wagner**, Yale University; **Marc Kirschner**, Harvard University; **Werner Callebaut**, Limburg University; **Eors Szathmary**, Collegium Budapest; **Alan Love**, University of Minnesota.

A central issue in making a new theory of evolution is how large a role natural selection, which has come to mean survival of the fittest, gets to play.

Natural selection was only part of Darwin's *Origin of Species* thinking. Yet through the years most biologists have mistakenly believed that evolution is natural selection.

A wave of scientists now questions natural selection's role, though fewer will publicly admit it. And with such a fundamental struggle underway, the hurling of slurs such as "looney Marxist hangover", "philosopher" (a scientist who can't get grants anymore), "crackpot", is hardly surprising.

When I asked esteemed Harvard evolutionary geneticist Richard Lewontin in a phone conversation what role natural selection plays in evolution, he said, "Natural selection occurs."

Lewontin thinks it's important to view the living world holistically. He says natural selection is not the only biological force operating on the composition of populations. And whatever the mechanism of passage of information from parent to offspring contributing to your formation, what natural selection addresses is "do you survive?"

In an aside, Lewontin noted natural selection's tie-in to capitalism, saying, "Well, that's where Darwin got the idea from, that's for sure. . . . He read the stock market every day . . . . How do you think he made a living?"

Stanley Salthe, a natural philosopher at Binghamton University with a PhD in zoology – who says he can't get published in the mainstream media with his views – largely agrees with Lewontin. But Salthe goes further. He told me the following:

> "Oh sure natural selection's been demonstrated . . . the interesting point, however, is that it has rarely if ever been demonstrated to have anything to do with evolution in the sense of long-term changes in populations. . . . Summing up we can see that the import of the Darwinian theory of evolution is just unexplainable caprice from top to bottom. What evolves is just what happened to happen."

Several months ago, Salthe hosted an intense email debate among leading evolutionary thinkers, which I was later let in on. It followed the appearance of an article by Rutgers University philosopher Jerry Fodor in the *London Review of Books* called "Why Pigs Don't Have Wings".

In the piece, Fodor – who told me he left MIT because he wanted to be closer to opera in New York – essentially argues that biologists increasingly see the central story of Darwin as wrong in a way that can't be repaired.

When I called Fodor to discuss the article, he joked that he was now in the Witness Protection Program because he'd been so besieged following the *LRB* piece. But we met for coffee anyway, on Darwin's birthday, as frothy snowflakes floated to ground around Lincoln Center. After a cappuccino or two, Fodor summed things up saying we've got to build a new theory and "all I'm wanting to argue is that whatever the story turns out to be, it's not going to be the selectionist story".

Fodor also told me that "you can't put this stuff in the press because it's an attack on the theory of natural selection" and "99.99% of the population have no idea what the theory of natural selection is".

Fodor noted in the *LRB* article that evolutionary investigators are looking to the "endogenous variables" for answers, which leaves plenty of room for interpretation. On that point there is considerable agreement.

But Richard Lewontin told me he resents evolutionary biology being "invaded by people like Jerry Fodor and others" as well as by some from within the field who don't really know the "mechanical details down to the last".

Evolutionary biologist and philosopher Massimo Pigliucci is also critical of Fodor for not seeing "the big picture". Pigliucci is a principal architect of the Altenberg 16 meeting as well as a participant. That rare combination – a consummate scientist with a sense of humor!

I met him one afternoon across the street from the New York Public Library during a break from his research. He had a birthday gift in one arm. Pigliucci says he enjoys life.

But while he thinks Fodor is "dead wrong" about natural selection becoming irrelevant to the theory of evolution, he does recognize the value philosophers, in general, bring to science. Several of the Altenberg 16 participants are, in fact, philosophers – including, of course, Pigliucci.

Pigliucci says philosophers have two roles to play in science. One is to keep scientists – who are focused on the details – honest by looking from a distance and asking the big questions: "Well, is the paradigm that you're working with, in fact, working? Is it useful? Could it be better?"

The second is as public intellectuals. He thinks some of the best responses he's seen against intelligent design and creationism, for instance, have been by philosophers. Pigliucci's philosophy web site "Rationally Speaking" carries the words of the Enlightenment's Marquis de Condorcet describing a public intellectual as one who devotes "him or herself to the tracking down of prejudices in the hiding places where priests, the schools, the government and all long-established institutions had gathered and protected them".

So what are those other engines of evolution that threaten to decommission natural selection – those "endogenous variables" – of which Jerry Fodor speaks in his now infamous "Why Pigs Don't Have Wings" article?

Pigliucci cites epigenetic inheritance as one of the mechanisms that Darwin knew nothing about. He says there is mounting empirical evidence to "suspect" there's a whole additional layer chemically on top of the genes that is inherited but is not DNA. Darwin, of course, did not even know of the existence of DNA.

Lewontin asks whether it's "Suspect or know?"

Nevertheless, these kinds of phenomena are part of what's loosely being called self-organization, in short a spontaneous organization of systems. Snowflakes, a drop of water, a hurricane are all such spontaneously organized examples. The self-organized systems grow more complex in form as a result of a process of attraction and repulsion.

So, coming up with a "sound" theory for form is one of the big challenges for the Altenberg 16.

Developmental biologist Stuart Kauffman is clearly one who thinks we must expand evolutionary theory. Kauffman, now head of the Biocomplexity and Informatics Institute at the University of Calgary, is known for his decades-long investigations into self-organization. He's been described by one evolutionary biologist as a "very creative man, try reading one of his books" who said in the next breath that "if he [Kauffman] really put an effort into understanding evolutionary biology – the basic theoretical framework that we have – I think he could have come a lot further".

Meanwhile, Kauffman's had a breathtaking career, beginning as a medical doctor, honored as a MacArthur Fellow (genius) and has worked with Nobel Prize winner Murray Gell-Mann at the Santa Fe Institute where he first studied self-organization. Looking at simple forms like the snowflake, he noted that its "delicate sixfold symmetry tells us that order can arise without the benefit of natural selection". Kauffman says natural selection is about competition for resources and snowflakes are not alive – they don't need it.

But he reminded me in our phone conversation that Darwin doesn't explain how life begins, "Darwin starts with life. He doesn't get you to life."

Thus the scramble at Altenberg for a new theory of evolution.

Kauffman also describes genes as "utterly dead". However, he says there are some genes that turn the rest of the genes and one another on and off. Certain chemical reactions happen. Enzymes are produced, etc. And that while we only have 25,000 to 30,000 genes, there are many combinations of activity.

Here's what he told me over the phone:

> "Well there's 25,000 genes, so each could be on or off. So there's 2x2x2x25,000 times. Well that's $2^{25,000th}$ Right? Which is something like $10^{7,000th}$. Okay? There's only $10^{80th}$ particles in the whole universe. Are you stunned?"

It's getting pretty staggering I told him. But there was more to come as he took me into his rugged landscapes theory –

hopping out of one lake into a mountain pass and flowing down a creek into another lake and then wiggling the mountains and changing where the lakes are – all to demonstrate that the cell and the organism are a very complicated set of processes activating and inhibiting one another. "It's really much broader than genes," he said.

Kauffman presents some of this in his new book *Reinventing the Sacred*. And natural selection is back in the equation

In his book *Investigations* (2000), Kauffman wrote that "self-organization mingles with natural selection in barely understood ways to yield the magnificence of our teeming biosphere". He said he's still there, but now thinks natural selection exists throughout the universe.

Stuart Pivar has been investigating self-organization in living forms but thinks natural selection is less relevant – and has paid the price for this on the blogosphere. Pivar's an extremely engaging man, trained as a chemist and engineer – a bit of a wizard – who loves old art. He was a long-time friend of Andy Warhol and a buddy of the late paleontologist Steve Gould, who continues to serve as an inspiration for Pivar's work.

Steve Gould's *Natural History* magazine editor Richard Milner, by the way, describes Gould as "a popular articulator of Darwinian evolution to a new generation, while privately, his creative and rebellious mind sought to move beyond it."

Milner, himself, is a Darwinian scholar and author of the *Encyclopedia of Evolution* and *Darwin's Universe*. He says Gould was intrigued with theories of how natural selection may act on levels beyond the individual (social groups, species), or at different phases of the life cycle (evolution-development), and how other embryological and evolutionary phenomena (heterochrony, neoteny) may influence or impact evolution. And he notes that "Gould took issue with those who used natural selection carelessly as a mantra, as in the evidence-free "just-so stories" concocted out of thin air by mentally lazy adaptationists".

Gould also famously rejected the reductionism of Richard Dawkins' "selfish gene" theory, Milner says further, and was well aware that there seemed to be a disconnect between the models of genes, DNA, and the development of individual plants or animals.

Says Milner:

> "Steve was one of the first evolutionary biologists, with Richard Lewontin, to publish the view that biology offered no plausible mechanism – a missing "theory of form," if you will – for how these genomic "blueprints" are followed in constructing phenotypes of living organisms."

I visited Stuart Pivar at his place just off New York's Central Park recently. It has the feel of a 19th century castle with interesting stuffed animals, rocks and other exotica, mixed in with important paintings and bronzes. Unlike most scientists I spoke with for this story, Pivar is not dependent on government grants to carry out his work.

Pivar says his theory is this. Body form is derived from the structure in the egg-cell membrane. And he handsomely illustrates in his book, *Engines of Evolution* [now titled *On the Origin of Form,* North Atlantic Books], how various species arise from the same basic structure, the Multi-torus, so-named by its discoverers – mathematicians, biologists Jockusch and Dress in 2003.

Pivar told me this structure was confirmed in recent years by Eric Davidson's identification of the sea urchin embryo as a dynamic torus, resembling a slow-moving elongated smoke ring – as in amoeboid motion.

If there's a lineage to his work, Pivar says it's rooted in Goethe, who observed that all life has a certain look to it – therefore it must be based on a form he called the "urform" – although Goethe never found the urform. Pivar's also been influenced by the 19th century scientist Wilhelm His, who made models using tubes of wax and pressed them to demonstrate how mechanical manipulation could generate the shape of the stomach, etc.

"The great D'Arcy Thompson was an inspiration as well," he said, citing Thompson's book *On Growth and Form* in which he described how every form in nature could be duplicated in the lab. Pivar said it's unfortunate Thompson never put the whole thing together to make a model, but that he has done just that.

He says he's shown that if you take a tubular form and you twist it this way or that way, you can generate the shape of anything in nature. He notes this is equivalent to the organization of chemistry by the periodic table. This twisting action is how tigers get stripes, butterflies wing patterns, as well as how the human embryo forms.

In *Engines of Evolution* [now *On the Origin of Form*], Pivar has published what he describes as "the blueprints" – the construction blueprints for the human body, frutfly, lobster, jellyfish – the scheme by which all nature forms.

Stan Salthe says he considers the theory of self-organization itself "up & coming" and thinks Pivar's idea is "reasonable".

Richard Lewontin told me the following:

> "I don't know what his [Pivar's] theory is but there's no question that the development of an egg is not dependent solely on the genes and nucleus, but on the structure of the egg as laid down to some extent. There are proteins that are there. There are non-genetic factors and I wouldn't be surprised if the actual structure of the cell

membrane had some influence on the successive divisions that occur."

However, Lewontin added that "it's one thing to say *some* effect than it is to say I have a theory that it's *all* there."

Pivar insists "It's all there."

Massimo Pigliucci does not consider Pivar's test with "wiggly water tubes" empirical evidence.

Pivar disagrees saying he presents a convincing model based on geometry and the animated drawings in his book but laments that he can't get serious circles to review his book. He attributes this reluctance to scientists being discouraged about taking a chance on ideas originating outside their peer group plus their dependence on government grants – which are tied in to support for natural selection.

Pivar's also a keen observer of some of the conflicts of interest tainting science. He accuses the National Academy of Sciences of excluding other approaches to evolution but natural selection in its recent book *Science, Education and Creationism.*

Richard Lewontin resigned from NAS over the issue of one branch of NAS accepting government funds for secret weapons programs.

Pivar is also critical of church and State influences in science education, like the *Astrobiology Primer* funded by NASA, whose editor is a priest.

Fodor goes further, he says, "Astrobiology doesn't exist. What are the laws?" [Fodor has updated his comment, saying he knows nothing about Astrobiology.]

Finally, Pivar thinks influential non-profits like the AAAS-affiliated National Center for Science Education, should not have religions represented on its board of directors. He is obliquely referring to NCSE's board member from the Church of Jesus Christ of Latter-day Saints-funded Brigham Young University.

Curiously, when I called Kevin Padian, president of NCSE's board of directors and a witness at the 2005 *Kitzmiller v. Dover* trial on intelligent design, to ask him about the evolution debate among scientists – he said, "On some things there is no debate." He then hung up.

Massimo Pigliucci finds it objectionable that "the study of forgiveness" is supported by the John Templeton Foundation, which funds the understanding of religion from a Christian view of God. [John Templeton Foundation was one of the sponsors of the March 2009 "Biological Evolution: Facts and Theories" gathering in Rome "under the High Patronage of the Pontifical Council for Culture."]

Pigliucci says the rationale of scientists who take this money is that it's hard to get grants, that they have to put their children through school, etc. "Well, yes – but there has to be a limit," he thinks.

As for educating the public about evolution, paleontologist Niles Eldredge, a co-author with Steve Gould of the punctuated equilibrium theory – which Eldredge reminded me was based on one of his early papers – says that increasingly scientists are being encouraged to include public outreach when asking for government grants.

Eldredge told me about the new journal that he and his son Gregory, a high school teacher in New York are publishing through *Springer* called: *Outreach and Education in Evolution.* It debuts in March and will feature peer-reviewed articles about evolution.

I also spoke with evolutionary biologist Michael Lynch at his lab at Indiana University to get his perspective on the evolution debate.

Lynch is the author of the recent book *The Origins of Genome Architecture.* He says it's hard enough just to be a molecular biologist or a cell biologist and that reaching out to communicate to other fields is a "daunting task". He doesn't know why there's a push for an Extended Evolutionary Synthesis and says, "Everyone's bantering around these terms complexity, evolvability, robustness, and arguing that we need a new theory to explain these; I don't see it."

Lynch thinks the big challenge is to connect evolution at the genome level with cell development and the larger phenotypic level.

I asked Richard Lewontin whether it was premature to put together a new synthesis. He said he wouldn't use the word "premature" and added, "Why would we want to do that? To say it's premature suggests that one of these days we have to. I don't know what we'll have to do in the future."

He continued:

"The so-called evolutionary synthesis – these are all very vague terms.... That's what I tried to say about Steve Gould is that scientists are always looking to find some theory or idea that they can push as something that nobody else ever thought of because that's the way they get their prestige.... they have an idea which will overturn our whole view of evolution because otherwise they're just workers in the factory, so to speak. And the factory was designed by Charles Darwin."

Clearly a new theory of evolution will impact all our lives. But how? Perhaps a global public broadcast of the Altenberg 16 proceeding is the answer to that question.

*****

## THE INVITE - "ALTENBERG 16" EVOLUTION SUMMIT

"We are writing to invite you to what we hope will be a major event to be hosted by the Konrad Lorenz Institute of Evolution and Cognition Research, in Altenberg, Austria (http://www.kli.ac.at/), on 10-13 July 2008. Our idea is nothing less than getting together a high-level group of biologists and philosophers to have a frank exchange of ideas about what, if anything, might a new Extended Evolutionary Synthesis look like.

As you know because you have been involved in this to some extent, for some time now there have been persistent rumors that the Modem Synthesis (MS) in evolutionary biology is incomplete, and may be about to be completed. Such suggestions have been received with skepticism by a number of biologists, including some of the very originators of the MS.

The challenge seems clear to us: how do we make sense, conceptually, of the astounding advances in biology since the 1940s, when the MS was taking shape? Not only we have witnessed the molecular revolution, from the discovery of the structure of DNA to the genomic era, we are also grappling with the increasing feeling – for example as reflected by an almost comical proliferation of "-omics," that we just don't have the theoretical and analytical tools necessary to make sense of the bewildering diversity and complexity of living organisms.

What is less clear is how much talk of an Extended Evolutionary Synthesis (EES) is actually going to coalesce into an organic conceptual structure capable of significantly augmenting the existing synthesis, while at the same time retaining the many key advances of Darwinism and neo-Darwinism – from population genetics theory to our still evolving understanding of the nature of species, to mention just two. The goal of the proposed symposium is, in fact, to accept the challenge and ask a number of prominent biologists and philosophers (see preliminary list of topics and contributors below) who have worked for an advancement of evolutionary theory exactly (or even approximately) what a meaningful EES would look like.

The central idea for the symposium is to have contributed papers on a range of conceptual issues that have not been addressed by (or at least are not an explicit part of) the MS, with the authors attempting not as much to give the latest technical update, but rather to provide an organic view of in what sense the new ideas can be said to extend the current scope of evolutionary theory. While it is of course impossible to be complete in such a bold survey of a rapidly changing field, we have put together a list of topics we think are crucial to an EES, and we are asking prominent scholars such as yourself to address such topics.

The goals of the workshop are two-fold: first, to bring a highly stimulating group of people together in Vienna to foster an open dialogue about the MS and the EES. Second, to produce a high-impact edited book (published by MIT Press), having the ambitious aim of providing a laboratory for ideas about what the EES might eventually look like. Since the intention is to have the book out for the Darwin anniversary year 2009, a prerequisite for accepting participation will be to agree to have a manuscript ready for the time of the workshop.

Both the workshop and the book are intended as tools for developing ideas, certainly not as finished products, yet, we think this very well could turn into a major stepping stone for the entire field of evolutionary biology.

We hope you will be joining us in Vienna next July. All travel and accomodation expenses will be paid for. If you are interested, please confirm your availability to us as soon as possible so that we may finalize plans for the workshop."

# Chapter 3

# Jerry Fodor and Stan Salthe Open the Evo Box

*"It works by selection of traits produced by random variations in the genes. That's essentially Darwin's hypothesis. I think not. . . . There's something wrong with the theory. It goes deep."* –**Jerry Fodor**

Rutgers philosopher Jerry Fodor's *London Review of Books* article "Why Pigs Don't Have Wings" (Oct. 2007) unleashed a serious online debate among evolutionary thinkers. The debate was hosted by philosopher Stan Salthe. I was fascinated and called Fodor to arrange an interview.

We met around Lincoln Center. Fodor loves opera and says his move back to New York from MIT was because of it.

There was a snowfall that day. Fodor and I trudged through it to Pain Quotidien for coffee. I wanted to talk about self-organization and he was bent on establishing that there were internal problems with the natural selection story of evolution. We were talking around the same thing. But a Jerry Fodor argument is something splendid to behold so I listened at fascinating length and taped, clatter and all.

Fodor said we've got to build a theory, but that in a certain sense – politics aside – it didn't matter who was right because "in 50 years we'll all be dead."

Maybe an argument there.

Fodor set out a way to look at the problem: "Here are the facts. And here are the prior theories. What do we have to change to deal with the data?"

What he did not establish was that the facts and prior theories are politically tainted. Who funded the finding of those facts and theories and promoted them?

Fodor described a couple of the theories on the table. One being Darwin's – that changes of inheritable properties are largely the effect of exogenous variables. There's an effective selection of who the predators are. The other is that there are effects we don't understand of endogenous variables and form.

He told me that if what is causing change is not selection, then maybe it is some laws of organization, but that *"basically I don't think anybody knows how evolution works."*

And then he said the following:

> "The heritable traits, features of biological organisms – complex or simple – change over time. They change as a function sometimes of variables or other god knows what. This would be true of the relation between any generation of the organism and the next generation and preceding generation.
>
> But the question that evolutionary theory is about, as opposed to questions about where did life start or something of that sort, the question of evolutionary theory is about when you get these changes in the inheritable structures of organisms – where do they come from? What are the controlling variables? It's not whether RNA comes before DNA – the basic question is: Are these changing shapes by environmental factors as in selection or are they shaped by some internal factors currently unknown?...

I say there's something wrong with the thesis that they're shaped by environmental factors. And so now there are various other alternatives."

As mentioned above, Fodor's story caught the attention of zoologist and natural philosopher Stan Salthe, who's also a visiting scholar at Binghamton University in New York. Salthe says his skepticism about natural selection has made him "poison" in some science circles. But he's demonstrated that he knows how to wage a pretty good frontline battle.

From October 16 - 24, 2007, Salthe ran an email chain on the Fodor "Pigs" story with some of the science elites throwing in their two cents.

I noticed that even a former philosopher-beau of mine had been indirectly dragged into the discussion. He does not have the best argument about evolution, however, so his comments do not follow.

Said **Stan Salthe**:

"Folks – There's not much new in this below, but, given Fodor's prominence, and the place of publication, I thought I would pass it on."

**Michael Ruse**, philosopher:

"In my opinion Fodor's piece is grotesquely and immorally irresponsible – he has done no homework on evolutionary theory – to say that natural selection did not shape the guppy and the fruitfly is ludicrous. Of course, every creationist in north america is salivating today – even though, they are the people who push adaptation more than even me! think that this one won't be used in the argument over what should be taught in schools?

Today, I am deeply ashamed to be a philosopher."

**Stuart Newman**, cell biologist:

"Fodor's piece seems pretty reasonable to me, in fact, kind of obvious. To say that organisms at any stage of evolution have only a limited array of condition-dependent inherent characteristics, or developmental pathways, and selection can do no more than choose among these, has nothing whatsoever to do with creationism."

**Michael Ruse**:

"You are very naive – it has everything to do with creationism – of course, to deny adaptationism is not to endorse creationism – but to write a piece slagging off natural selection in that way, is to give a piece of candy to the creationists – I am sure that duane gish has already incorporated this into his talks.

of course natural selection has to work on an array of given things, but this is not to deny selection – especially not through fodor's silly arguments about analogies – and certainly not adaptationism.

the point of course is that fodor did not simply write a technical piece on adaptation – he wrote a piece flamboyantly denying selection. In today's climate, where we have just had two ultra right supreme court justices appointed, I think his behavior is somewhere between stupid and wicked."

**Bob O'Hara**, mathematician:

"Hmm. For me it could do with a bit more substance, explaining what's wrong with Fodor's essay. At the moment it comes across as if you don't like the guy, but not really why we should take your side."

**Guy Hoelzer**, organismal and molecular biologist:

"Hi David [Sloan Wilson] and all...

First, I should say that I found Fodor's article to be

an interesting read... It is typical, and appropriate IMHO, that proponents of emerging alternative paradigms shout loudly and exaggerate their claims. How else are they to get the attention of the masses holding fast to conventional wisdom? [That was a rhetorical question. There may be other ways, but loud exaggeration is one natural option.] Convention is very difficult to overturn and I think that science would be well served if the defenders of convention were more tolerant (open-minded) of possible new paradigms. It is a good idea to hold challengers to higher standards, but I don't think we should spitefully relish in putting them down, which is all too easy to accomplish when we are preaching to the choir of convention.

Believe me, I know that you of all people appreciate the importance and difficulty of fighting conventional wisdom, and I am reacting less to your draft than to the set of messages impugning Fodor's article. As weak and potentially flawed as the presented arguments may be, I think the essence of Fodor's article is a call for extending the scientific conversation about evolutionary processes, especially those involved with adaptation, beyond the traditional bounds of Darwinism. The long-term benefit of listening to the challengers, even though some (many?) of them might waste our time, could be worthwhile. Let's not add too much to the distracting noise of fighting with heretics."

**Robert J. Richards**, science historian:

"Dear Will [Provine],

You suggest Fodor is confused by Darwin's two representations of natural selection. But this supposes he actually read the *Origin*, which the evidence is strongly against..."

**Elliot Sober**, philosopher:

"Dear David [Sloan Wilson],

I did want to comment on your last comment where you say

[**Wilson**]: "I end by calling attention to this passage of Fodor's essay, which no one has commented upon yet: "Why is it so hard to be good? Why is it so hard to be happy? One thing, at least, has been pretty widely agreed; we can't expect much help from science. Science is about facts, not norms. It might tell us how we are, but it couldn't tell us what is wrong with how we are. There couldn't be a science of the human condition."

Does anyone other than myself find this passage objectionable? Note that it targets science as a whole, not evolution or adaptationism. Apart from its formulaic rendering of the naturalistic fallacy as a settled issue, it ignores the fact that implementing any value or norm requires knowledge about the facts of the world. This passage tells me that Fodor does indeed deserve the title of "secular creationist." We wouldn't be surprised at this kind of anti-scientism coming from a postmodernist, and we should recognize it for what it is in Fodor."

[**Sober**]: "Fodor was speaking loosely here but I'm sure he would not deny that "implementing any value or norm requires knowledge about the facts of the world." What he is talking about is that statements about what our values ought to be are not something that science describes. Science can tell you that doing X will get you Y, but it doesn't tell you that you ought to strive to obtain Y. I agree with this. This isn't anti-scientism or postmodernism, but a recognition that the naturalistic fallacy is a fallacy."

**Stan Salthe**:

"As a kind of postmodernist myself, let me say that this statement shows some ignorance of THAT perspective. The postmodern perspective, insofar as it contacts science, is that the practice of science is a socially constituted discourse, from the production of, *e.g.*, microscopes, to the form of scientific writings (*i.e.*, third person, global present tense) to, at the other end, who sponsors what kind of science and why.

**All of these need inquiry as to how they may be biasing the scientific enterprise.**

... [S]imply stated my critique of the concept of natural selection...is that it is suspect because it so snuggly fits into our culture's obsession with competition (from ping-pong through banking to warfare) that it is an idea that cannot be resisted, heavily freighted with taking it for grantedness... Michael's [Ruse] reaction to the Fodor paper is quite typical of that of neoDarwinians to any challenge to their exaggeration of one aspect of the original Darwinian project (as by Charlie himself!), and to any challenge to their narrow-minded powerful hegemony.

What is occurring now is that outsiders have indeed been drawn into this discourse, by way generally of the complexity/self-organization discourse (from which Piattelli-Palmarini [Fodor's book co-author] comes), as found, *e.g.*, at Santa Fe. David's [Sloan Wilson] view that only experts in biology can deal with this theory is quite at odds with the views of systems science, semiotics and complexity discourse generally. We can ALL look over the logic of this dominant driving idea (competition) of our culture. These ideas have jumped out of biology some time ago, (back) into economics, into psychology, etc. Everyone is now explicitly instead of only implicitly involved... ALL, politics!...

The cat is out of the bag! Darwinism is everywhere, and it is the logic of its positions that are under scrutiny by anyone who is trained to think; *i.e.*, by an intellectual. The complexity folks are already deeply involved."

**Stuart Newman**:

"Phenotypic plasticity is the primitive condition of all biological systems. Thus, even if adaptation can be demonstrated in some modern forms, *e.g.*, the beak size and shape in finches, this is hardly paradigmatic of how macroevolutionary change took place."

**David Sloan Wilson**, biologist:

"Another example of the need to fight for the middle ground concerns the recent atheist attacks on religion by Dennett, Dawkins, Hitchens and others. In this case the problem is not a denial of natural selection but other departures from responsible scholarship to portray religion as bad, bad, bad in every respect. This includes not only poor scholarship with respect to theology, which Jack might appreciate given his comment about creationism, but extremely poor scholarship with respect to scientific understanding of religion, from an evolutionary or any other perspective. It doesn't matter that these authors have distinguished reputations, any more than a creationist with a PhD. They need to be roundly criticized on the basis of their current effort…

This is not repression. It is fighting to prevent responsible scholarship and scientific inquiry from becoming a carnival."

**Jack Maze**, botanist:

"My God, I can't believe I just read this! Are we to take this to mean that Fodor must be stopped or that his views, many of which I find interesting, must be prevented from being disseminated? What I read here is an erudite version of the polemics coming from G.W. Bush and I find it offensive in the extreme.

If, on the other hand, the fear is that Fodor will give aid and comfort to the creationists then I would recommend you ignore him and attack the creationists at their weak point, they rely on bad theology. ID, as a God of the gaps argument, leads inevitably to idolatry, and violation of the First Commandment, and the God depicted by Young Earth Creationists is one that is unacceptable to thinking monotheists.

Please do not include me among your friends."

**Guy Hoelzer**:

"David [Sloan Wilson]

... It is the assymetry of 'piling on' that I think interferes with progress toward Truth. This is why I try to find and emphasize the thought-provoking aspects of an article like Fodor's. Rigor should be our criterion for evaluating new ideas, including whether they are coherent, but I don't think we need to be so quick to pass judgement that we squash the good ones before they are ready to compete."

**Peter Corning**, complexity theorist:

"So what about Fodor's core point that nonadaptive changes from various sources may also be important causes of evolutionary change. Well, of course. This has been part of the dialogue among evolutionists ever since the publication of D'Arcy Thompson's great treatise *On Growth and Form* (1917). His work was further elaborated by C.H. Waddington and Bernard Rench in the 1950s, by Eldredge and Go [Gould], Ho and Saunders (among others), and, more recently by Stu Kauffman, Stan Salthe and his colleagues, and others of the dynamical systems, structuralists, evo-devo and complexity schools. Yes, causation in evolution is immensely complicated and variegated.

Kauffman has been perhaps the most visible and vocal (and contentious) among the current generation of theorists in proclaiming that much of what we see in the evolutionary process may be characterized as "order for free" – biological organization that arose independently of natural selection. However, Kauffman has since retreated from this salient, and for good reason..."

# Chapter 4

# Theory of Form to Center Stage

*"Well, it's very unfortunate that you bought into this. There are very few people in evolutionary biology who take Pigliucci seriously, and Fodor, Pivar, et al. are literally unknowns, providing no evidence that they've read a single bit of the mainstream evolutionary biology literature. Of the pictures of the Altenberg group, I'm sure that most are unaware of your posting, as I know that several of them are mainstream evolutionary biologists, with no axe to grind with the framework in which most of us are working. We are seeing the field of evolutionary biology being increasingly trivialized and threatened by quacks, and although these folks are by no means creationists, they are every bit as dangerous. All areas of inquiry have limitations, but there is no justification for criticizing a field until one has put in the effort to understand it."*
– **Mike Lynch**, Distinguished Professor, Dept. of Biology, Indiana University

*March 25, 2008*
*2:02 pm NZ*

IT'S PRIVATE. That is now the word from organizers as to whether or not the public can listen-in to the conversation of 16 scientists meeting to remix the theory of evolution at Altenberg, Austria in July. So while Konrad Lorenz Institute where the symposium will take place may not exactly resemble the sacred pond of Emperor Augustus where priests read the entrails of eels and advised what was to befall Rome – and evolutionary science is nowhere near as primitive – there is still public concern about the emergence of evo high priests, as reflected in a substantial response to my recent series of evolution stories.

Evolutionary biologist, **Massimo Pigliucci**, a co-architect of the KLI "major event", emailed me saying "somebody could do a sociological study of science and fringe science just based on the developments from our interview."

As a rising star on the lecture circuit, Pigliucci knows there is much public interest in a new theory of evolution. Why then must the Altenberg proceedings remain closed?

Here's my abstract:

> *The public says a reformulation of the theory of evolution is bigger than the Altenberg 16 – no matter how brilliant the individuals may be. People think a remix of evolution is high priority and say they have a right to know, since scientists are publicly funded. They want open discussion and say such information should not be locked down for future book and DVD sales. There is too much of that kind of industry already corrupting science.*

Following is feedback on the evolution stories – what developmental biologist **Stuart Kauffman** might consider part of the "ceaseless creativity" of the universe.

I was happy to see **Richard Leakey's** web site pick up the Dawkins story. It was actually Richard Leakey, then director of the Nairobi museum, who secured the plane for my flight into Olduvai in 1980 to talk with his mother, the late paleoanthropologist Mary Leakey.

A few of the science sites that picked up the Altenberg story include: ***Genome Technology Online, Geoscience Research Institute, Nature Publishing Group***, and ***Society of Systematic Biologists***.

The Altenberg articles have been discussed online in at least a half dozen languages: **Latvian**, **Indonesian**, **Portuguese**, **Dutch**, **Italian** and **Spanish**.

And they've been posted to religious sites, such as the **Harekrishnas** and **Church of Jesus Christ of Latter-day Saints**.

Intelligent Designers are apparently some of the most vigorous bloggers on evolution, and Paul Nelson's column on the Altenberg story for ***Uncommon Descent*** generated 206 comments.

***Democratic Underground***, always a fierce battleground, took up the debate, excerpting natural philosopher and zoologist **Stan Salthe's** comments regarding natural selection, *i.e.*, "Summarizing we can see the import of Darwinian theory of evolution is just unexplainable caprice from top to bottom. What evolves is just what happened to happen."

**Michael Purugganan**, a professor of genomics and one of the Altenberg 16 scientists, emailed from his lab at NYU to say (in jest?), "I'll have appropriate buttons or pins made that say Altenberg 16... they will be distributed at the meeting."

And Yale professor of ecology and evolution **Gunter Wagner**, another of the Altenberg 16 – who I was unable to reach for comment for the original story – wrote "Hey that is great! This is one button I will keep!"

But there have been grumblings elsewhere in the scientific community about who was left out of the Altenberg 16 invite. The name most mentioned as crucial to a successful redo of evolutionary theory and missing from the Altenberg 16 list is that of Harvard evolutionary geneticist **Richard Lewontin**.

Lewontin told me he would not go, however. "No. No way." Aside from any personal objection to the content of the conference, there may be another reason Lewontin is not attending. According to one admirer, Lewontin doesn't like to fly.

One of the other giants I interviewed for the first Altenberg piece – paleontologist **Niles Eldredge**, originator of the punctuated equilibrium theory with Steve Gould – was apparently also not invited.

Eldredge told me he only "knows of" Massimo Pigliucci, and actually described himself as "a very conventional evolutionary biologist". "I disappoint people sometimes," he said.

Evolution's star public intellectual, **Richard Dawkins**, might have been too much of an anchor for the Altenberg 16 discussions. He wasn't invited, as he hinted during our recent Q&A at Barnes & Noble in Tribeca:

> "You've been taken in by the rhetoric... You asked the question: Have I been invited? I'm sorry to say I get invited to lots of things and I literally can't remember whether I was invited to this particular one or not."
>
> But it's being viewed as a "major event", I told him.
>
> "By whom I wonder?", he jabbed.

There was a disgruntled posting by "Jim C" from Melbourne on ***Dawkins.net*** saying, "We may indeed know more about the nature of DNA and the complex details of its operations, but it is the height of intellectual arrogance to describe the evolutionary theory moving away from a 'population genetic-centered view'."

With noticeable rivalry in the scientific community then, one of the qualifications for the Altenberg 16 no doubt must have been congeniality.

Massimo Pigliucci's colleague from Stony Brook, evolutionary biologist **Doug Futuyma**, for instance, is not on the list either.

University of Toronto biochemist **Larry Moran**, who runs a popular web site called ***Sandwalk***, which considers itself the rival to SEED blogger **PZ Myers'** ***Pharyngula***, asked me: "Why was Doug Futuyma not invited?"

Niles Eldredge shared the following with me about Futuyma's treatment of him and Steve Gould in his book:

> "If you open Doug Futuyma's book – the guy at Stony Brook who's probably one of the most famous evolutionary biologists in the country now if for no other reason than he wrote that widely-read text book – you're not going to find that Steve Gould and I get a very good shake in that book. And you're not going to find I don't think an extended discussion of self-organization, if it's even mentioned."

Maybe Futuyma's lack of interest in self-organization – a subject which will be explored at the Altenberg 16 meeting – kept him off the list.

That Lewontin, Eldredge, Dawkins and Futuyma will not be a part of the so-called Extended Evolutionary Synthesis discussion in Austria does diminish the event. But there's plenty of fresh talent out there as well. And Massimo Pigliucci and **Gerd Müller** have presumably found some.

**Stuart Newman**, a professor of cell biology and anatomy at New York Medical College and board member of KLI, is presenting his theory of the origin of form at Altenberg – the current synthesis, remixed 70 years ago, has been without a theory of form. Newman says he doesn't know half the people who are going to be there and that scientists really need a forum in which to interact with like-minded scientists.

Newman added, "These groups are kind of carefully chosen so as not to create fights."

The real disconcerting issue, though, is the public being shut out. The ***New York Times*** has now asked to attend after reading my Altenberg story. They too have been turned away. Not the first time the *Times* has noted my work, incidentally: "New York Times Fesses Up to Another Rip Off". http://counterpunch.org/mazur11182004.html

Larry Moran at *Sandwalk* agreed with my suggestion that the conference somehow be made public. Moran linked my posting of the Altenberg 16 invite, to which one reader (**The Monkeyman**) responded:

> "It is a shame that this and other conferences that hold interest to many budding scientists who cannot get the money to attend... will not be streamed or recorded and put up for download."

Newman says "private" maybe evokes the wrong kind of image. He says he agrees that a lot of science is publicly funded and he does feel a responsibility to make his research accessible to the public. But he also thinks the event may not be designed for "prime time".

Says Newman, "If I were to take my undigested ideas and make them accessible to the public, I would lose my reputation very fast because I probably have a lot of stupid ideas also."

Gunter Wagner commented on Moran's web site in response to the Altenberg piece, that he didn't like being lumped together with the likes of philosopher **Jerry Fodor** and **Stuart Pivar**.

Fodor wrote the now infamous "Why Pigs Don't Have Wings" story in the *London Review of Books* arguing that the central story of the theory of evolution is wrong in a way that can't be repaired, which led to a fierce online exchange among leading evolutionary thinkers hosted by Stan Salthe.

Pivar is the independent scientist whose work has been skewered on the blogosphere for not being a complete theory of evolution.

But despite the controversy, look for more commentary from Jerry Fodor on evolution without adaptation. He's taken a year off from teaching as State of New Jersey philosopher at Rutgers University to write a book with **Massimo Piattelli-Palmarini**, a professor at the University of Arizona, who Fodor says is handling the biology. Fodor also says he has tenure and is not worried about fallout.

But Salthe also told me he doesn't like being lumped together with Fodor and Pivar, because he *has* published papers in the field. On the other hand, this doesn't seem to make a difference in his attempts to contact A-16's Gunter Wagner, who Salthe says won't respond. Salthe says he thinks his view of natural selection makes him "poison".

**Fodor says the careers of most evolutionary scientists are tied up with natural selection.**

Pigliucci defends the natural selection turf. He told me this:

> "So to say that natural selection is out of the picture seems to me to discard literally thousands and thousands of empirical papers. And I don't think anybody can afford to do that, let alone Fodor, who is apparently not familiar with that literature."

But Salthe says you can't dismiss the censorship going on in the evo debate. He recently sent me his correspondence with the Neo-Darwinian journal *TREE* (*Trends in Ecology and Evolution*) in which he asked them to publish his letter arguing to "save the phenomenon of convergent evolution even if it seems inconvenient" to the Darwinian perspective on organic evolution. Salthe was responding to an article *TREE* published suggesting the concept of convergent evolution be eliminated based on a totally genetic analysis. *TREE* refused to publish Salthe's letter.

Convergent evolution he says happens when different species become similar without involving "similar genetic representation". Examples of this he notes are old world vultures evolving from hawks and new world vultures evolving from storks...

PZ Myers/*Pharyngula,* who Richard Dawkins quoted adoringly at Barnes & Noble last weekend, again trashed Stuart Pivar's work in his blog after reading the Altenberg 16 story, but liked the fact that I'd interviewed Richard Lewontin.

Pivar says he did take the advice of NASA mineralogist Robert Hazen and early on approached mainstream evolution publishers. He has been repeatedly rejected he says, but continues to fight on, making the point that he's the only one with a model.

Pivar recently offered a research grant to Massimo Pigliucci and his lab to study his book on self-organization following an exchange of emails with Pigliucci over several months.

Pigliucci said he considered the gesture "bribery" and refused the offer, adding that he does not share Pivar's enthusiasm about his theory of form.

**Richard Milner, Steve Gould's** former editor and author of *Darwin's Universe,* told me that Darwin, while terrified of controversy himself, was no shrinking violet in sending out his disciples, Thomas Huxley ("Darwin's bulldog") *et al.,* to proselytize for him. Milner said Darwin kept a list of those he converted to his theory and that Huxley was shut down by the police for holding meetings on Sundays where he'd present his science lectures as sermons, calling them "Lay Sermons".

I also received an email from a microbiologist named Anders in Denmark, who said Pivar's concept was not sound science – which I passed on to Pivar to address. Pivar said he emailed Anders, who had not read his book, and told him he was sending a copy to him for review.

Pivar says he welcomed a recent email with constructive comment from Stuart Newman on the subject of form.

Newman told me this in a phone conversation about the theory of form he'll be presenting at the Altenberg meeting:

> "The idea is that when multicellularity first emerged [a half billion years ago] the products of a number of genes that existed in single-celled organisms, and had evolved for single cell purposes, began to mobilize physical forces and physical processes that are characteristic of materials at a larger ('meso') scale. For instance, single-celled organisms or individual cells can secrete molecules, but it doesn't really have any consequence for the structure of the organism. But if it's in a multicellular context, then secreted molecules can form gradients that provide patterns… It's a simple form of self-organization.
>
> This can help explain the burst of animal evolution that happened in the Cambrian explosion. Consider, in a more complex example, two very distinct physical processes that existed in the single-celled world, the production of secreted molecules and intracellular biochemical oscillations. When they found themselves together in a multicellular-scale structure, their

> combined effect was to make segmentation all-but-inevitable. In fact, we know that modern-day embryos, including those of humans, still use these ancient "generic" physical processes to form their segmented backbones.
>
> When multicellularity emerged, new physics was brought to bear on the formation of organisms. It's not that these physical processes didn't exist before multicellularity. They just didn't pertain to the development of organisms before...
>
> At the point when the modern animal body plans first emerged [half a billion years ago] just about all of the genes that are used in modern organisms to make embryos were already there. They had evolved in the single-celled world but they weren't doing embryogenesis. What did it take to get them to do embryogenesis? It took a change in scale. What led that change in scale is that, possibly due to alterations in external conditions, cells became sticky. And once they became sticky, you had multicellular organisms, and mobilization of the self-organizing physical processes of mesoscale materials."

Newman also said people are working on pre-biotic evolution, but nobody would be presenting a paper on the subject at Altenberg in July.

Fast forwarding a half billion years, Newman's got a very interesting article in the March 2008 issue of *Capitalism, Nature, Socialism,* called "Evolution: The Public's Problem and the Scientists'" in which he observes that:

> "The nearly exclusive focus on genes to account for biological change at the levels of both individual development and large-scale

> evolution, like the cash nexus of market economies, collapses quality into quantity, life into symbol."

Speaking of symbols, **Ben Stein's** ***Expelled*** film people contacted me "to get involved" with the project after reading the Altenberg story. Curiously, they take a pro-war position, and I have not responded to them.

**Sam Smith**, the inimitable editor of ***Undernews***, the online publication of ***Progressive Review***, excerpted and linked the Altenberg story with pinups of the various evo thinkers not invited to the Altenberg 16 meeting (including the late Andy Warhol), plus that of Massimo Pigliucci. The *Undernews* comments were growing tense last time I looked.

Financial guru **Catherine Austin Fitts**, a former Assistant Secretary of Housing in the Bush I administration, included the Altenberg 16 invite story on her ***Solari.com*** "Top Picks" list.

Canada's ***Financial Post*** and ***National Post*** linked the Dawkins story.

New York's ***Daily News*** posted the Altenberg piece. And ***Fox Television News*** carried the Altenberg 16 story on Sean Hannity's forum.

So the public is craving evolution enlightenment. But **Massimo Pigliucci** has said he will keep the lid on the Altenberg proceedings regardless and that he and European co-organizer **Gerd Müller** will comment for the group only after returning home.

"You're saying that you're ruling out even a panel in Altenberg at the end of the conference?" – I asked Pigliucci.

"You mean on the order of: Doctor, how's the patient doing?" – he responded. "No, the meeting is private."

# Chapter 5
# The Two Stus
# Stuart Kauffman – Peace, Love and Complexity

"I have worked on self organization for years and love it," Stuart Kauffman emailed me from the University of Calgary's Institute for Biocomplexity and Informatics, which he founded. What's more Kauffman pinpointed exactly where we are in the quest to unlock the secrets of evolution during a phone conversation on Valentine's Day 2008.

Said Kauffman: **"There are people spouting off as if we know the answer. We don't know the answer."**

Nevertheless, Kauffman has moved on to tackling the "ceaseless creativity" of the universe in his book, *Reinventing the Sacred*. He described the book to me in an email as "VERY radical scientifically... and dangerous culturally," adding "I'm waiting for Jaweh to hit me with a thunderbolt, shades of Thor."

Following our phone conversation, he sent me a second email clarifying that the book "aims to try to create safe spiritual space across all our traditions". He elaborated:

> "At a time when, with no insult to Catholics, the Pope states that the Church is the only true church, while Jews (like me), Muslims, and the Dali Lama are condemned to Hell, when Left Behind is a video game urging children to kill those who do not embrace Jesus for the Rapture, with Islamic extremists strapping bombs and killing people, we just have to try."

# Stuart Kauffman: Rethink Evolution, Self-Organization is Real

*May 5, 2008*
*1:21 pm NZ*

In his new book, *Reinventing the Sacred,* legendary complexity pioneer Stuart Kauffman continues to challenge the view of most biologists that natural selection is the only source of order. However, Kauffman is more charitable than hundreds of other evolutionary scientists (non-creationists) who contend that natural selection is politics, not science, and that we are in a quagmire because of staggering commerical investment in a Darwinian industry built on an inadequate theory.

True to his research roots in self-organization, Kauffman says life is not based on the replication of DNA and RNA. He also questions whether biology can be reduced to physics, writing that lovers walking along the Seine are not just particles in motion.

He thinks the biosphere constructs itself using sunlight and free energy and that the universe is "ceaselessly creative." And because the future is not really predictable, Kauffman (writing from the Canadian Rockies) recommends we all calm down, remix science with the ancient Greek model of "the good life, well lived," and treat ALL in our global culture as sacred.

Stuart Kauffman draws on 40 years of work for the book, from his investigation of snowflakes to "coherence-decoherence" of the conscious mind.

Kauffman tackles evolution of the economy as well. Yes, it's ceaselessly creative.

He comes clean in a chapter called "Broken Bones" revealing he has advised the US Joint Chiefs of Staff on asymmetric warfare and terrorism (Kauffman's been a consultant to Los Alamos too). He notes that "all sought to prevent war," but that "history shows us that war is often excused by a trumped-up atrocity or threatened atrocity."

I telephoned Stuart Kauffman because I wanted to discuss self-organization, an area he trailblazed in the 1960s at New Mexico's Santa Fe Institute.

Kauffman began his career as a medical doctor, has been honored as a MacArthur Fellow, a Marshall Scholar and awarded the Gold Medal of the Academia Lincea Rome. He is a founder of the University of Calgary's Biocomplexity and Informatics Institute and is currently an adjunct professor in the university's philosophy department.

His three previous books are: *The Origins of Order, At Home in the Universe*: *The Search for the Laws of Self-Organization and Complexity*, and *Investigations*.

Stuart Kauffman spoke passionately about self-organization for much of the 45 minutes of our pre-scheduled talk, taking me on a shamanic flight through his rugged landscapes theory and to the edge of chaos throughout the universe. Then after several pleas (they grew loud) from his handler that my allotted time was up – the vision ended – and Kauffman agreed to move on to his next appointment.

Excerpts from our interview follow in **Appendix A**.

# Stuart Newman – The Chess Master

"He's a one to watch in the unfolding evolution discourse," I was advised regarding New York Medical College cell biologist Stuart Newman. The money was riding on him, so to speak, at Altenberg to provide the convincing theory of form. Coincidentally, Stuart Kauffman was on Newman's PhD committee.

Stuart Newman is the classic evolutionary scientist, dedicated to his lab and the nurture of his students, like NYMC grad student Ramray Bhat, who has his name on Newman's DPM papers as co-author.

Newman has told me he thinks the public has a right to share in scientific discovery and that he tries to make his papers publicly accessible. "I do care," he said.

Newman enjoys penning critical professional articles on science and culture as well. And he's testified before Congress when asked.

A former student said Newman keeps a Spanish masterpiece on his computer screen for inspiration. (it's actually "The Apotheosis of Henry IV and the Proclamation of the Regency of Marie de Medicis on May 14, 1610".)

It's unclear whether Stuart Newman actually plays chess, but he's a grandmaster at checking the opposition, as the following emails reveal:

> "**4/2/2008** Thank you for your kind words about the paper, Suzan... please note, in contrast to other workers in the field – Stuart Kauffman, for instance, but particularly Eric Davidson – we

actually de-emphasize the specific role of transcription factors in the generation of form. See the section of the paper that deals with "DTFs."

**4/23/2008** I'm disappointed in Dawkins. His response is neither public nor understanding.

**5/8/2008** Thanks for this. It sent me to the Sherman paper, which I had not seen. (I have attached it.) I'm not too impressed. The only causal agent is cryptic "programs." It doesn't explain anything, but exhibits a "generative combinatorics" which is also a feature of our model. I'm also disappointed in Chomsky for endorsing it without probing deeper. It's not very serious of him.

Piattelli-Palmarini has an instinct for what's wrong with neo-Darwinism, but he seems to want to fix it with a not-too-coherent eclecticism [See the Piattelli-Palmarini Q&A, Appendix D]. Eric Davidson's descriptive biology is very good, but his conceptual framework for evolution is pure neo-Darwinism, and as different as can be from ours. Stuart Pivar is a brilliant eccentric with an aesthetically based Platonistic obsession. His developmental details are almost all incorrect and I don't think his ideas will contribute at all to an eventual synthesis.

**5/11/2008** Thanks for your great efforts to raise the level of discourse (as they say in science studies circles). Of course Chomsky is right about the "Thompson-Turing approach, with its roots in rational morphology." I'm back to being a fan…

**5/12/2008** He [Chomsky] thinks Sherman is working in the Thompson-Turing tradition, which is not the case. We are, however. Nonetheless, I would suggest not pressing him on this. He will not be an important participant in this discourse, and there is no need to irritate him, which seems on the verge of happening.

**6/17/2008** Interesting article. I agree with L-D-F [Antonio Lima-de-Faria] about Darwinism. But self-assembly is not the answer. Just a temporal version of preformationism: http://en.wikipedia.org/wiki/Preformationism; *i.e.,* there is nothing new under the sun.

> "This order is also patent in the cellular shaping of a living organism. At present it is known that the pattern of an embryo is decided by a large collection of small and large RNAs, i.e., pure atomic processes, which have the 'road map' that decides the cellular pathways."

This could not be more wrong.

**6/4/2008** Hi Suzan, I just should tell you that every time you mention Stuart Pivar in one of your articles it cuts your credibility among actual scientists to a small percent of what it would otherwise be. Though mentioning my work in the same context does not do me much harm, since it leads to dismissal of your views, it does me no good either."

Newman also told me that he objected to the title "The New Charles Darwin," but not to "The New Master of Evolution", adding, "Let's hope it's true." Eight minutes later he followed up saying: "I meant the theory part, of course, not the master part."

# Stuart Newman, The New Master of Evolution?

*April 8, 2008*
*11:37 am NZ*

> *"A molecule of water does not form itself into waves and vortices. However, a mass consisting of trillions of water molecules does. You don't need new substances. It's just a matter of changing the scale and bringing new physical processes into play.*
>
> *So if you have cells with genes that evolved for single-cell functions, and you put the cells together into clusters, then those clusters of cells have physical properties – including self-organizing properties – that individual cells never had. And then those cell clusters can make 40 or 50 different kinds of forms very rapidly by virtue of the physics mobilized by the existing gene products in the changed scale. They don't need a lot of genetic changes to go from one form to another. And the forms produced in this way can become locked in later by natural selection."* – **Stuart Newman**

On his recent PBS television show *One On One,* John McLaughlin quizzed two religion authors about whether the idea of the Virgin birth in the oral accounts of the Bible had a reference in biology, with someone mentioning parthenogenesis – asexual reproduction. But could those early storytellers have actually been feeling their way around the idea of orthogenesis, *i.e.,* spontaneous birth?

The idea of orthogenesis was particularly popular with 19th century scientists, such as Ernst Haeckel and Jean-Baptiste Lamarck but was then abandoned. Orthogenesis has regained credibility.

Paleontologists Niles Eldredge and Steve Gould suggested in their pivotal paper on punctuated equilibrium, for instance, that there was a deficiency of transitional forms. Further scientific investigation has revealed why, and now the 150-year old so-called theory of evolution – which has never had a coherent theory of form – may be about to get one.

It was thrilling and somewhat humbling to read cell biologist Stuart Newman's hypothesis of the evolutionary triumph of life as it self-organized by the time of the Cambrian explosion roughly half a billion years ago – using what he calls toolkit genes – making a balletic leap from single-celled highly plastic organisms into multicellularity. All 35 or so animal phyla formed around this time, as evidenced by the fossil rock.

The toolkit genes, some of which act to mobilize basic physical forces and processes, thereby becoming what Newman calls **DPMs (dynamical patterning modules)** – along with others, the **DTFs (developmental transcription factors)** – performed an almost hallucinatory dance on the page as I read through his paper about a pattern language enabling organisms about a millimeter in size to body-build – cavities, layers of tissue, segments, extremities, primitive hearts and even eyes.

And while the DPMs got the DTFs to arrange the matter of cell types and functions – Newman says it was the DPMs that really provided most of the pizzazz.

**Interestingly, Newman says this part of evolutionary history turns the Darwinian theory upside down in the sense that natural selection is not central.**

Stuart A. Newman is currently Professor of Cell Biology and Anatomy at New York Medical College in Valhalla, New York where he teaches and directs a lab.

He has collaborated with University of Vienna theoretical biologist Gerd Müller, University of Missouri biological physicist Gabor Forgacs as well as Ramray Bhat, on aspects of his DPM hypothesis. He has also co-authored the textbook *Biological Physics of the Developing Embryo* (Cambridge University Press) with Gabor Forgacs, and with Gerd Müller co-edited *Origination of Organismal Form: Beyond the Gene in Developmental and Evolutionary Biology* (MIT Press), a volume about the origination of body form during Ediacaran and early Cambrian periods, also contributing a few chapters to it.

Newman's A.B. degree is from Columbia University and his PhD in chemical physics from the University of Chicago where he also did post doctoral studies in theoretical biology, as well as at the School of Biological Sciences, University of Sussex. Newman's been a visiting professor at Pasteur Institute, Paris; Commissariat a l'Energie Atomique-Saclay and the Indian Institute of Science, Bangalore.

But Stuart Newman is also a public intellectual. Through the years he has commented on social and cultural issues affecting science. He chose not to stand on the sidelines regarding the ethical issues surrounding human genetic and bioengineering, for instance.

In 1997 along with Jeremy Rifkin, then president of the Foundation on Economic Trends, Newman attempted to patent a part human, part animal chimera to highlight the dangers of the commercialization and industrialization of organisms, which he fears will ultimately include humans.

In 2002 he testified before the US Senate saying:

> "Since my student days I have also been concerned with the uses to which scientific research is put. Having become convinced that scientists, who are beneficiaries of public resources [Newman has been the recipient of NSF and NIH grants for 30 years], have a deep responsibility to anticipate what lies down the road in their own fields and to serve as a resource

> for the public on the complex issues around applications of scientific research, I joined with other scientists, social scientists, women's rights advocates and environmentalists, to found the Council for Responsible Genetics in the late 1970s."

Newman has also weighed-in on politics' impact on science in capitalism and socialism, noting how political interference can indeed warp research and deprive society of its benefits. Newman wrote recently in "Evolution: The Publics' Problem and the Scientists" appearing in *Capitalism, Nature, Socialism*:

> "The Soviet doctrine of Lysenkoism represented by ideological distortion of evolutionary biology that may be thought of as generic to top-down socialism... the genetic determinist ideology that it both rejected and gave life to... comports well with the worldview of advanced capitalism."

But getting back to Stuart Newman's hypothesis – **what exactly did the DPMs (dynamical patterning modules) – I count nine – look like singly? And moving in combination?**

It seems that if these DPMs were physical-genetic modules, as Newman says in the paper, there ought to be a way to visualize them.

Did some appear as kind of wobbly electric neon discharges? And others maybe resemble the old Camel smoke ring that once wafted through Times Square?

Newman does provide a schematic drawing at the end of the paper showing the effects of various DPMs acting singly and in combination. And there is a table listing the DPMs with the molecule each is linked to, its physical principles and its role in evolution-development.

He also says what DPMs were made of. They were gene products (proteins) but also networks of physical processes that shaped and patterned organisms using adhesion, polarization, viscoelasticity, *etc.* DPM elements also combined with one another to make oscillations and segmentation, as well as morphogens from secreted multicellular molecules.

The morphogens then traveled through the organism to create organismal forms. Newman says the survival of cell types later became a matter for selection.

There were also certain signaling pathways, like the Notch pathway, he says, that factored in. Notch was established several billion years ago before multicellularity occurred, as was the Wnt pathway, for example. Wnt in conjunction with the Frizzled family receptors of signals made cells polarize.

Newman explains how spots and patterns were generated by morphogens. Lateral-acting inhibitors and reaction-diffusion systems came into play. He says skeletal elements, bones, feathers and hair were created this way.

He also indicates that while some **DTFs (developmental transcription factors)** did exist in single-celled organisms for the purpose of cell contractility and light sensitivity, other DTFs – those participating in segmentation and tissue identity – did not exist in single-celled organisms. And he notes that combinations of DPMs and DTFs enabled segmentation, skeletogenesis, eyes to form and heart-like structures to develop.

Stuart Newman concludes that it was the multicellular state that led to the appearance of developmental mechanisms, and that transcription factors in regulating cell fates did not play a starring role in the pattern language at the time of the Cambrian explosion.

He says he's not comfortable with the title "The New Charles Darwin", though. Well, how about "The New Master of Evolution"?

See Appendix B: "Stuart Newman's "High Tea"" and Appendix C: "The Enlightening Ramray Bhat".

# Chapter 6

# The Two Massimos

# Massimo Pigliucci – Evolution & Flamboyance?

Massimo Pigliucci is a man on the move. Yes he's got three PhDs, and as the *New York Times* style pages noted last year in covering his wedding, he's had three of those too.

Pigliucci gave me roughly 45 minutes or so of his time on the evolution debate in a meeting in February, when he again reminded me about Altenberg. There was virtually nothing personal about our conversation. And once I began writing about the Altenberg conference, which Pigliucci coordinated with scientists on this side of the Atlantic, he began shutting down communication.

He kept his distance at the May 2008 gathering at the 23rd Street library, where the Secular Humanist Society of New York arranged for him to speak about "paradigm shift" and his book on junk science. In fact, in responding to my questions about natural selection, there was no indication he'd ever even met me before. Certainly Richard Dawkins would have remembered! (Chapter 10, "Richard Dawkins Renounces Darwinism as Religion")

Indeed, as one of Pigliucci's colleagues mentioned, Massimo's flamboyance gets in the way. Pigliucci had a couple of awkward moments at the Secular Humanist Society talk, for instance.

For some peculiar reason, he referred to a quote from "feminist philosopher" Sandra Harding in the *New York Times* about the "husband as scientist forcing mother nature to his wishes," with Pigliucci adding that he didn't feel he was doing that in his lab.

Made me think about the fact that there's only one woman in the Altenberg 16.

Pigliucci next attempted a quip about a French philosopher, who he said a lot of postmodernists have claimed as one of their inspirations. The only thing he had in common with the philosopher, he said, was an interest in cooking and that "if you mention the name [inaudible] I'm likely to reach for my gun."

Pigliucci will need some talented speechwriters if he's to overtake Dawkins on the evolution stage. As he described the Sokal hoax, for example, I found myself trying to visualize philosopher Paul Boghossian instead reading from his sublime piece on Sokal in the *Times Literary Supplement*.

Author and historian Richard Milner then stole the show from the back of the library with a story about Piltdown, as Pigliucci grew increasingly red-faced.

The second the event ended, Pigliucci fled. I took the opportunity to say hello to his wife, whose previous life as an adventurer and a director of the International Rescue Committee impressed me.

# Massimo Piattelli-Palmarini – Evoluzione senza Adattamento

Massimo Piattelli-Palmarini is intrigued by origin of form. He's the professor of cognitive science at the University of Arizona who's co-writing *What Darwin Got Wrong* with philosopher Jerry Fodor. Piattelli-Palmarini is handling the biology for the book on evolution without adaptation. He's got a PhD in physics from the University of Rome.

Through the years Piattelli-Palmarini has been a visiting professor at Harvard, MIT, the University of Bologna and the College de France in Paris. He spent eight years as principal research scientist at MIT's Center for Cognitive Science. It was at MIT that Piattelli-Palmarini first met Jerry Fodor and linguist and beloved activist Noam Chomsky, eventually collaborating with both on books.

He's also served as Director, Florence Center for the History and Philosophy of Science and Director, Royaumont Center for a Science of Man, Paris.

Piattelli-Palmarini is the author of a half dozen or so books, notably *Inevitable Illusions* and *Ritrattino di Kant ad Uso di Mio Figlio (Portrait Kant to Use My Child)*, for which he received Italy's Premio Tevere for non-fiction.

He was also awarded the Accademia d'Abruzzo, Premio Il Rosore d'Oro for his work as a public science intellectual. And he's a regular science contributor to the Italian newspaper *Il Corriere della Sera.*

In describing Charles Darwin's theory of natural selection, Piattelli-Palmarini tells the story about a boy asking his father why rocks fall to the ground and his father answering that when the Earth was young some rocks floated away from the planet, some suspended in air and eventually floated away, and some fell to the ground. The remaining ones on the ground fell to the ground.

More of Piattelli-Palmarini's wit is in evidence in this email to me about my interview with Stuart Kauffman:

> **5/6/2008** "Interesting interview (by the way it's George Boole, not Bool). Some points really are perplexing. How can it be that "There's no molecule in Ghadiri's system" if there are proteins? That's VERY strange. Also, by the way, the German Nobelist chemist Manfred Eigen developed auto-catalytic cycles of RNA and proto-proteins back in the early Seventies. Maybe he is cited by Kauffman in the book, but the interview gives the wrong idea that he and his colleagues have been the first.
>
> Monod and Jacob's model is NOT one of reciprocal switching. If the Stu gene can turn the Sue gene on or off, it's not overwhelmingly the case that the Sue gene can turn the Stu gene on or off. Regulatory genes are distinct from structural genes. Moreover I really do not follow Stu in his anti-reductionist humanism (lovers strolling on the banks of the Seine is a sheer distractor from the main issues). Of course they are not molecules, but who ever claimed they are? He has a bit of hubris that I do not like too much. Interesting anyway. Massimo"

Full interview, **Appendix D**: "Piattelli-Palmarini: Ostracism without Natural Selection".

# Chapter 7
# The One and Only Richard Lewontin

**Richard Lewontin** – *Harvard Geneticist*

*Phone Conversation February 25, 2008*

**Suzan Mazur**: You say here ["The Triumph of Stephen Jay Gould", *New York Review of Books* 2/14/2008] regarding Steve Gould that "while he could not hope to smash utterly the Darwinian icon, he certainly wanted to put a noticeable crack in it." Can you say anything more about that? Do you feel essentially that Steve Gould wanted to demolish natural selection?

**Richard Lewontin**: No no no no. That's not really the point. I guess I was being a little archist there. The point is that Steve, like lots of ambitious scientists, wanted to do better than the famous people who came before him. So if he can show that there's some aspect in which he could improve on Darwin, why he does it. That's all.

He doesn't want it to be an icon. But he's not opposed to natural selection. Certainly not. He never was opposed to natural selection.

**Suzan Mazur**: Is it your opinion that natural selection is here to stay?

**Richard Lewontin**: Natural selection occurs. The problem for the biologist is that natural selection is not the only biological force operating on the composition of populations. There are random forces because after all population is only finite in size. And even if there's no natural selection, everybody does not have exactly two children. Every couple doesn't exactly have two children to replace it. And there's randomness in which sperm fertilizes which egg. So things change for that random reason. And things change because species go extinct. Nobody knows why any particular species ever went extinct. But every species goes extinct.

**Suzan Mazur**: Now is it your feeling that the current focus on the endogenous variables – which is actually what Jerry Fodor's talking about – that there's a whole lot of attention to this subject and that a lot of people are looking at what's loosely being called self-organization. I was wondering if you think that this aspect of the science, this look at self-organization, is going to play a much larger role and natural selection may be less of an influence on evolution.

**Richard Lewontin**: I don't know what Jerry Fodor even means by self-organization. I really don't.

**Suzan Mazur**: I don't think he used the word *per se*.

**Richard Lewontin**: What's self-organization? I mean there's development. An organism develops from egg to adult. And its development from egg to adult depends on an interaction between what genes it's got inside itself and what environment it grows up in...

I don't really understand what it's [self-organization's] all about. Look, every higher organism – forget about bacteria for the moment – goes through an egg and a sperm. So what are all these endogenous variables? I don't know what they're all about.

An organism develops and there are some constraints on development. And the development of one part of an organism influences the development of another part of an organism. But that's all within a generation. What has it got to do with passing information on from one generation to another? I'm sorry I don't understand.

You were formed by the fertilization of one of your mother's eggs and by one of your father's sperm. And we know a lot about spermogenesis and oogenesis. There are maternally inherited characteristics. There certainly are. But we don't have any information that those…

**Suzan Mazur**: Can I ask you another question regarding the Modern Synthesis? There's a movement to create what's being called an Extended Evolutionary Synthesis because the feeling is that the Modern Synthesis based on Darwin's theory of evolution has been great as far as it goes – it's worked for 70 years – but there have been so many theories coming up recently that a major change is needed.

**Richard Lewontin**: Well, why?

I'm sorry you're really talking to somebody who doesn't… I mean I'm probably a bad person to talk to because I don't pay much attention to all these theories. Theories are cheap. And people make a living by making theories.

**Suzan Mazur**: But there's a major meeting on this coming up in July in Vienna – on this Extended Evolutionary Synthesis.

**Richard Lewontin**: Well, that might be, but I still don't know what the content is.

**Suzan Mazur**: The feeling is that there are some phenomena we've discovered recently that don't fit with the existing theory.

**Richard Lewontin**: That might be.

**Suzan Mazur**: With the concept of the Modern Synthesis.

**Richard Lewontin**: But I don't really know what those phenomena are. Can you give me an example?

**Suzan Mazur**: Yes. The idea of epigenetic inheritance, for instance. One argument is that "we're beginning to have mounting empirical evidence to suspect that there's a whole additional layer of inheritance. Not just the genes."

**Richard Lewontin**: Suspect or know?

**Suzan Mazur**: There's mounting empirical evidence.

**Richard Lewontin**: Well there are mitochondria. There are a variety of things and there is the passage...

**Suzan Mazur**: It's not in the genes. It's not DNA. So what are they?

**Richard Lewontin**: But what has that got to do with decommissioning natural selection? Whatever the mechanism of passage of information is from parent to offspring – some of it's environmental, some of it's genetic, and so on. But natural selection doesn't ask that question. It asks the question – ok, once this information has contributed to your formation, do you survive or are you likely to have more offspring and a higher probability of survivorship and so on. That doesn't change the equation of natural selection. And the claims that evolutionary trends...

**Suzan Mazur**: I think the people meeting to discuss this are not looking to toss out natural selection. But it's not clear how much a role these other processes come into play. So there's a push to...

**Richard Lewontin**: But to come into play into itself cause progressive, I mean successive – not progressive because that has other meanings – but successive evolutionary changes so an organism will get bigger and bigger or smarter and smarter or redder and redder over generations or simply that they provide another mechanism.

**Suzan Mazur**: Another layer.

**Richard Lewontin**: I don't know what a layer is, another mechanism of inheritance.

**Suzan Mazur**: Well, I mean that's what they're investigating.

**Richard Lewontin**: Well, we know there are other mechanisms in inheritance beside genic. We know that there is cultural inheritance. We know there is internal inheritance due to the cytoplasm, etc. Yes, we know all those things. But the thing I'm asking you is why that represents a directional force or a changing force along…

**Suzan Mazur**: Because the feeling is that the Modern Synthesis says that all evolutionary change is gradual. That there are no jumps. This really changes the equation with a focus on…

**Richard Lewontin**: I don't understand the relationship between jumps and these things. If these things are from individual parent to individual offspring, then before a population can become as a totality changed – this is the old problem of Lamarck – every individual in the population has to be changed. But that means that the change has to pass through the population increasing in frequency generation after generation. There's confusion here, it seems to me, between the message of passage of information and whether or not more and more individuals of a particular kind will make up the population.

**Suzan Mazur**: So you think it's short-lived the change?

**Richard Lewontin**: Well it's one generation. It's not a question of short-lived. It's a question – look, you have to understand the basic form of our reproduction is such that whatever information passes to me from my parents does not pass to your parents. That the chief point about Darwinian evolution is that there's variation in the population. There's some mechanism of passing information about those variants from parent to offspring. And the way evolution works is, that a larger and larger fraction of the population for some reason has a particular characteristic.

Unless you had an infectious form of information passage in which every time I touch somebody their inheritance would be changed. It isn't basically different from the Darwinian model. After all Darwin knew nothing about genes. He never heard of genes.

As you describe it to me there seems to be some confusion between the two issues. I don't see why the passage of information from parent to offspring by cytoplasmic factors like mitochondria leads to a jump in the population.

**Suzan Mazur**: You mean because it's only short-lived.

**Richard Lewontin**: It's one generation. But even if it weren't short-lived, even if every individual in that line of inheritance acquired that, you still have a million five hundred thousand other lines of inheritance of the population and they don't acquire it from you unless it's infectious. And evolution is not a change in an individual line, it's a change in a whole population of individuals who make up the species.

**Suzan Mazur**: So you think it's premature to be talking about an Extended Evolutionary Synthesis?

**Richard Lewontin**: I wouldn't use the word premature. Why would we want to do that? To say it's premature suggests that one of these days we will have to. I don't know what we'll have to do in the future.

The so-called Evolutionary Synthesis. These are all very vague terms. And you have to understand as a writer about science – and that's what I tried to say about Steve Gould – is that scientists are always looking to find some theory or idea that they can push as something that nobody else ever thought of because that's the way they get their prestige. Do new and different things.

So our field – especially evolutionary biology – has been invaded by people like Jerry Fodor and others. People who don't really know the mechanical details down to the last and also people from within the field. And Steve did a certain amount of this. **[People] who are always looking to suggest that they have an idea which will overturn our whole view of evolution because otherwise they're just workers in the factory, so to speak. And the factory was designed by Charles Darwin.**

You want to become famous, you make a big noise about some of your latest theory.

**Suzan Mazur**: So you wouldn't characterize Steve Gould as someone embracing self-organization then.

**Richard Lewontin**: **I don't even know what self-organization is.** Self-organization is a very confusing term to me.

**Suzan Mazur**: To everyone. But they all seem to working in it.

**Richard Lewontin**: Yeah. I understand. **What I'm trying to tell you is – for me at least that's not science.** If I can't do experiments that relate directly to a very clear and logical hypothesis – I don't want to talk about self-organization. I don't know what it means.

**Suzan Mazur**: Have you heard about the theory that Stuart Pivar has been discussing.

**Richard Lewontin**: Stuart who?

**Suzan Mazur**: Stuart Pivar. He's a chemist and mechanical engineer.

**Richard Lewontin**: Sorry, I don't know.

**Suzan Mazur**: He has a theory that form originates from a pattern in the egg cell membrane.

**Richard Lewontin**: I don't know what his theory is but there's no question that the development of an egg is not dependent solely on the genes and nucleus, but depends on the structure of the egg as laid down to some extent. There are proteins that are there. There are non-genetic factors and I wouldn't be surprised if the actual structure of the cell membrane had some influence on the successive divisions that occur.

But it's one thing to say some effect than it is to say I have a theory that it's all there.

Look, I think you should take – every time somebody says I have a new theory, I suggest you turn off your hearing aid.

**Suzan Mazur**: Now there are some arrows being flung in your direction regarding your economic – the economic view that you have.

**Richard Lewontin**: What have they got to do with evolution?

**Suzan Mazur**: I was just wondering if you consider that unfair. I mean this whole discussion of capitalism, etc.

**Richard Lewontin**: Well what about it? What does it have to do with evolution?

**Suzan Mazur**: Well some people feel that natural selection has everything to do with capitalism.

**Richard Lewontin**: Well that's where Darwin got the idea from. That's for sure.

**Suzan Mazur**: Oh. Really.

**Richard Lewontin**: Sure. How do you think he made a living.

**Suzan Mazur**: But you're embracing natural selection.

**Richard Lewontin**: What's the relationship between that and the way society organizes itself? Is there any law that says we have to organize our society according to natural selection just cause Darwin got the idea about evolution or about natural selection from the society in which he lived?

**Suzan Mazur**: Right…

**Richard Lewontin: Well, we don't *have* to organize human society "Nature, red in tooth and claw [Herbert Spencer's famous reference to survival of the fittest]". No. We don't have to.**

**Suzan Mazur**: So you think we should be looking at other possibilities.

**Richard Lewontin**: Of the organization of society. But what's that got to do with how evolution occurs?

**Suzan Mazur**: Ok.

**Richard Lewontin**: They're two different things.

**Suzan Mazur**: Well, I'm glad you're feeling better. And I really do appreciate your comments.

**Richard Lewontin**: But you understand I have – it's pretty obvious from this conversation – I have very little sympathy with people who just spin out new theories all the time. We're loaded with them. And the problem is that they're attractive to write about in newspapers and magazines. And this review of books and that review of books. But they're not science. They're just made up out of people's heads. They're what Steve and I used to call "just so" stories.

**Suzan Mazur**: Just so?

**Richard Lewontin**: Kipling's just so story. How the camel got its hump. People feel free to make it up out of their heads. And I'm sorry, I don't feel free to make it up out of my head.

**Suzan Mazur**: So you won't be attending this event in Vienna?

**Richard Lewontin**: No. No way.

**Suzan Mazur**: Well thank you so much.

**Richard Lewontin**: Sorry to not be more helpful.

**Suzan Mazur**: No no no. You were really helpful. I very much appreciate it.

**Richard Lewontin**: I'm coming from someplace else than these people.

**Suzan Mazur**: Thank you so much.

**Richard Lewontin**: Ok, bye-bye.

# Chapter 8

# Knight of the North Star: Antonio Lima-de-Faria, Autoevolution

*June 17, 2008*
*4:50 pm NZ*

### A. Lima-De-Faria: Autoevolution, Atoms to Humans

> "For a long time the experimental studies carried out on elementary particles, chemical elements and minerals did not reveal that they had an evolution of their own. This picture has been drastically changed in recent years by the finding that the elementary particles, the chemical elements and the minerals have each had an autonomous evolution. Hence, three separate evolutions occurred before the biological one emerged.
>
> Evolution starts when the universe is born. And this is not a loose process since the elementary particles already show specific ancestors and specific rules of evolution. Later the chemical elements of the periodic table also display an ordered and well-defined evolution. Still later, the minerals also undergo an evolution of their own. These three separate evolutions preceded the biological one.
>
> It is this new evidence on the physico-chemical phenomena that now allows biological evolution to be placed in a completely different context. Most important is the fact that, since biological evolution was anteceded by these three levels, it became a prisoner of these previous evolutions. The laws and rules that they followed created the frame from which biological evolution could not and cannot depart.

> I may confess that I have a certain aversion for creating new terms. However, in this case I felt compelled to coin the word autoevolution. It describes the transformation phenomenon which is inherent to the construction of matter and energy. This consequently produced and canalized the emergence of forms and functions." – **Antonio Lima-de-Faria**, *Evolution without Selection. Form and Function by Autoevolution*, 1988, Elsevier, New York (page 18)

> "This was published 20 years ago. Nothing like the test of time to assess the validity of a scientific statement. The evolution of elementary particles (J.A. Pons *et al.* 2001), that of the chemical elements (E.R. Scerri 2007) and that of the minerals (B. Sandstrom 2006) have been corroborated since then by a vast body of research establishing the three evolutions on a firm basis. This evidence has revealed that they followed similar paths and carried their mark intact into the biological levels." – **A. Lima-de-Faria**, June 2008

Sweden's king decorated molecular cytogeneticist Antonio Lima-de-Faria "Knight of the Order of the North Star" for his outstanding experimental work, which elucidated the molecular organization of the chromosome and its evolutionary path.

Lima-de-Faria says this contribution was the product of work at the laboratory bench and at the light and electron microscopes for 13 hours a day, without weekends and vacations, for 30 years (which resulted in over 192 scientific papers).

The molecular analysis, using radioisotopes and other methods, made him realize that the chromosome, and the gene, arrived quite late in evolution and that their function and transformation were guided by the atomic processes which preceded them during mineral formation.

He opens two chapters in his book *Molecular Evolution and Organization of the Chromosome,* 1983, Elsevier, New York (pages 3 - 25) noting "The gene is not so important" and "The chromosome is not such an important structure either". And he points out that the chromosome, due to its rigid molecular organization, evaded both selection and randomness (pages 129 - 162).

**Lima-de-Faria does not consider Charles Darwin's 1859 idea of natural selection – survival of the fittest – a theory.** He writes in his classic book, *Evolution without Selection. Form and Function by Autoevolution,* that Darwinism and the neo-Darwinian synthesis, last dusted off 70 years ago, actually hinder discovery of the mechanism of evolution.

Lima-de-Faria says life "has no beginning; it is a process inherent to the structure of the universe."

Ever look at the swirls of frost on a window pane and discover similar patterns in the curled shoots of a fern? Or maybe wonder why an insect is leaf-shaped? Or why brain coral and the human brain look alike?

Antonio Lima-de-Faria concludes from these and other observations that seeming coincidental patterns arise in living organisms because they have the same atoms as minerals, with the symmetries of minerals "transferred intact to the cell and organism level".

He says "the human body is built on the atomic plan of a twin crystal" with the actual symmetry of the crystal "decided by the electronic properties of the constituent atoms." It is not an exact symmetry, however – physicists think that total symmetries happened only briefly at the time of the Big Bang.

What's more, Lima-de-Faria says we've now collected enough data that it's possible to construct a preliminary periodic table in biology like that in chemistry, one that accounts for form and function of organisms occurring at regular intervals in evolution. In fact, he first presented a working biological periodicity table in his book *Biological Periodicity. Its Molecular Mechanism and Evolutionary Implications*, JAI Press, U.S.A. 1995 (page 283), along with a discussion of various functions and structures, such as the occurrence of the placenta, flight, vision, penis, bioluminescence and high mental ability.

He says the best example is flight, which emerged in insects but in none of the other many invertebrate groups. Then it appeared suddenly in pterosaurs (flying reptiles which were not directly related to insects. Later it developed in birds (thought to have originated from flightless reptiles since pterosaurs had died out).

Mammals radiated into many families, but Lima-de-Faria notes that only bats were able to fly and they had no immediate affiliation with birds.

Another example he cites is the formation of the penis, which does not occur in most fish, amphibians and birds. However, lower invertebrates, such as flatworms, have developed almost as complex an organ as that of the human male penis – with a seminal vesicle, a prostate gland and ejaculatory duct. The penis structure has appeared at various times in evolution, he says, being well developed in snails, barnacles and mammals.

Lima-de-Faria thinks it's significant that the periodic recurrence of these structures and functions is not related to the general environment of organism complexity. What he says lies at the base of this biological mimicry is molecular and atomic mimicry, well established in minerals and proteins.

Antonio Lima-de-Faria is Emeritus Professor of Molecular Cytogenetics at Lund University in Sweden. He has a PhD in genetics from Lund and completed his undergraduate studies in biology at the University of Lisbon in his native Portugal. A. Lima-de-Faria founded and served as Director, Institute of Molecular Cytogenetics, Lund University for almost 20 years.

He has been a Fellow of The Rockefeller Foundation, International Atomic Energy Agency and the Institute for Cancer Research in Philadelphia. He's also been a visiting professor or visiting scientist to a half dozen research institutes and universities, including Duke and Cornell in the US, the Max Planck Institut fur Meeresbiologie in Germany, the Centre de Recherches sur les Macromolecules in France, the National Institute of Genetics in Japan and the Institute of Animal Genetics at Edinburgh University, UK.

Lima-de-Faria has been honored with appointments, prizes and medals from various countries (aside from Swedish knighthood). Among these are Portugal's Great Official Order of Santiago and Sweden's Gold Medal for State Service. He is a member of five scientific academies.

The six books he's published include: *Evolution without Selection. Form and Function by Autoevolution,* 1988; *Handbook of Molecular Cytology,* 1969; *Molecular Evolution and Organization of the Chromosome,* 1983; *Biological Periodicity, Its Molecular Mechanism and Evolutionary Implications,* 1995; *One Hundred Years of Chromosome Research and What Remains to be Learned,* 2004; and his latest book *Praise of Chromosome "Folly": Confessions of an Untamed Molecular Structure,* from World Scientific, Singapore.

As one would expect from a Swedish knight, Antonio Lima-de-Faria prefers personal communication. Our conversation has taken place by telephone and through air mail packages.

He shares some of his extraordinary insights in the interview that follows.

**Suzan Mazur**: You've called natural selection "the opium of the biologist for over 100 years," saying it is an abstract concept, and as such it can't be measured and poured into a vial – and that the term natural selection should be removed from evolution vocabulary because it is a hindrance to the discovery of the mechanism of evolution.

You acknowledge that natural selection exists but say it has nothing to do with the basic mechanism of biological transformation, which is based on physico-chemical and mineral layers of evolution. So why are most biologists and textbooks and scientific academies still embracing natural selection?

**A. Lima-de-Faria: Selection is a political not a scientific concept. At the time of Darwin it fitted perfectly the expanding colonialism of Victorian England. At present, Darwinism has been equated with evolution in an effort to convert it into the ideological arm of globalization. For this reason it will remain a powerful force until this system will be superseded by a more humanitarian form of economic development.**

Nothing could be better than selection because it can "explain" equally well a given situation or its opposite state. This is why there are as many Darwinist interpretations as there are authors. The result is total confusion.

Moreover, Darwinism starts from the wrong end of evolution. *The Origin of Species* is about a terminal process in biological transformation, thus it cannot give an answer to a phenomenon that started billions of years ago.

Everybody knows that selection occurs in nature, but the chromosome and the cell circumvent its effect by many molecular mechanisms. Among these are: DNA repair (which corrects atomic disturbances arising during DNA replication), RNA surveillance (which disposes of aberrant RNAs ensuring the formation of nondefective proteins) and the well known immune system that counteracts and destroys foreign agents adverse to the cell and the organism.

Moreover, if a given structure or function has disappeared, it may resurface due to the occurrence of biological periodicity.

**Suzan Mazur**: You've stated that we've never, in fact, had a theory of evolution and that Charles Darwin's idea of natural selection was not a theory. Is there any point then in making an Extended Evolutionary Synthesis?

**A. Lima-de-Faria**: There is no place for compromise. That is what most people have done before and continue to do. The long procession of criticisms of Darwinism and "new theories" that were supposed to substitute it were easily silenced. The main reason is that most of them, if not all were compromises, in which selection was still given some role in an undefined form.

The other main reason was that no physico-chemical alternative was produced. There was no sufficient evidence on the self-assembly of molecules and cells and on the molecular mechanisms that established the coherence of chromosome organization. The result was that they were easily dispensed with.

Since you are from New York, may I take the liberty to mention that in 1964 when I was a visiting professor at Duke University (Anatomy Department), I was invited to occupy the Chair of Genetics at Columbia University, a position which I declined.

A previous colleague at this university was Thomas Morgan who developed Drosophila genetics (Nobel laureate 1933). He stated that science was 99% perspiration and 1% inspiration. I must confess that I have not even pretended to have that 1%.

Most of us are aware of the limitations of our knowledge and are only compelled to draw the logical conclusions of the results that we have accumulated. This is why I never called my novel concept of autoevolution a "new theory". Theories, in advanced sciences, such as chemistry and physics, are based on a coherent body of knowledge that allows predictions.

No prediction seems to be possible at present concerning biological transformation. Darwin could not tell, and no one can tell today what species will come after humans, sparrows or lilies. Since I always abhorred abstract models and "armchair theories", which abound in the literature, I looked for a physico-chemical mechanism that may not "explain" evolution but may elucidate its origin and dynamics.

**Suzan Mazur**: Should scientific papers be dispensable like art? If it's bad art, tear it up and make a masterpiece. The current argument of science elites, though, is that they can't just throw away thousands of papers and move in another direction – even if it's the right direction.

**A. Lima-de-Faria**: In science you never throw away any well established result; you only elucidate better a given phenomenon as new technologies allow a deeper insight into the underlying mechanism. Independently of Darwinism, the study of evolution – from paleontology to DNA sequencing – has led to an impressive body of knowledge that is a formidable achievement and that will stand the test of time.

Few phenomena in biology are so well established as evolution. The great body of research in molecular biology and allied fields is totally independent of the Hollywood vision of Darwinism, and cannot be shaken by the fashions of the day.

**Suzan Mazur**: The irony is scientists are the best funded intellectuals and in exchange for government grants agree to ensure the public welfare – yet funds are being squandered on the outdated Darwinian and neo-Darwinian approach to evolution. To what do you attribute this?

**A. Lima-de-Faria**: Grants are awarded by your colleagues who sit in Research Councils and Foundations. Most of us, in any establishment, tend to be conservative and to follow what is called the paradigm. **This creates a cycle of submission**. All courage that you may have in life is never enough. One prerequisite in the scientific endeavor is that you are so stupid that you never understand what your colleagues say.

**Suzan Mazur**: You describe biological evolution as a "prisoner of the rules and principles guiding the construction of matter and energy". You also say biological evolution is terminal. What happens next?

**Lima-de-Faria**: Obviously, no one knows what will happen next. Instead we need to concentrate on what is well established at present, that may give us a key to the order evident in biological transformation.

**The origin of form and function must be sought in the process of self-assembly**. This is not an abstraction but a permanent event which has been demonstrated to take place at the level of elementary particles (by physicists), atoms (chemists), macromolecules (biochemists) and cells (biologists).

**The experimental results have been available for the last 35 years but have been ignored or silenced to avoid creating cracks in an edifice based on randomness and selection.**

**Self-assembly is a term coined by biochemists to describe the spontaneous aggregation of multimeric biological structures involving formation of weak chemical bonds between surfaces with complementary shapes. Isolated subunits can spontaneously assemble in a test tube into the final structure, a process that is inevitable and automatic.**

At the first stage quarks and antiquarks united into mesons and other particles. Later protons, neutrons and electrons self-assembled into atoms, and at a further stage atoms assembled into crystals. At the protein level, the 12 dissociated units of aspartase transcarbamoylase are able to reconstitute the active enzyme.

Nucleosomes are the essential components of chromosomes. When their purified DNA and histones are isolated, they spontaneously rebuild the chromosome thread.

As one reaches the viruses, their RNA and proteins, as is the case in tobacco mosaic virus, will spontaneously reassemble producing infectious particles. Living organisms make the same feat.

The self-assembly of the dispersed amoeboid cells of *Dictyostelium* results in a complete slime mold. The animal *Hydra* can self-assemble from its dispersed cells which are highly complex.

As one reaches the organs of mammals, dispersed liver cells, kept in culture, can self-assemble into a functional liver and dispersed cells from human organs, like those of skin and capillaries are also able to self-assemble into these tissues and organs. There is no need for any form of disguised obscurantism or vitalism.

Experiments have demonstrated that the resulting cellular order is due to the production of specific molecules that are recognized by the cell's surface proteins. This evidence was a critical component in allowing me to propose a mechanism of evolution based on a rigid internal atomic organization. (*Biological Periodicity. Its Molecular Mechanism and Evolutionary Implications*, 1995).

**Suzan Mazur**: Are we close to recreating biological evolution?

**A. Lima-de-Faria: Actually, in the case of some organisms, this recreation has occurred long ago. The best established example is that of a plant species found in nature that was obtained by experimental means.**

**It was the leading Swedish geneticist A. Muntzing who, in 1930 working with *Galeopsis* (hemp nettle), crossed two different species and by doubling their chromosome number obtained *Galeopsis tetrahit* which occurred spontaneously in nature. This work is not even mentioned in the *Encyclopedia of Evolution* (2 volumes, M. Pagel, Editor, 2002).**

**Muntzing's work does not support the dogmatism of Darwinism based on random mutation and selection. In the case of *Galeopsis*, no successive random mutations were needed, only crosses between two different species, and no selection was involved. The spontaneous doubling of the chromosomes resulted in the new species.**

**Silence is the strongest weapon. The disregard for science's ethical principles is widespread.**

**Suzan Mazur**: In *Evolution without Selection,* you note that the nucleus has no obvious ancestor and no one seems to know where it comes from. Do you have further thoughts on this?

**A. Lima-de-Faira**: As far as I know, there is no further information on its ancestry. What is remarkable is that the nuclear envelope furnishes the best example of the unfailing power of self-assembly. **At every cell division, and there have been an astounding number since the dawn of the cell, the nuclear envelope has self-assembled with a tremendous precision ensuring the formation of normal daughter cells**. It is one of the most striking examples among the mechanisms responsible for the maintenance of biological order.

This order is also patent in the cellular shaping of a living organism. At present it is known that the pattern of an embryo is decided by a large collection of small and large RNAs, *i.e.*, pure atomic processes, which have the 'road map' that decides the cellular pathways.

**Everything is rigidly ordered in the universe, otherwise there could not be variation because disintegration would occur**. In other words, paradoxically, a frame consisting of molecular constraints is an obligatory condition for variation to occur, otherwise any form, or function, would have soon decomposed into an unrecognizable pattern.

**A human's body shape has been repeated for over one million years. Without molecular internal coherence, an amorphous mass would have emerged at every new generation. However, this order cannot be total, otherwise there would be no place for transformation.**

**The key is movement – permanent atomic movement – that obliges alternatives (called errors by the less initiated)**. Significant is that these alternatives are not of all kinds (*e.g.*, during the replication of DNA its bases are substituted by identical bases or by base analogs).

**Only those molecules or atoms that partly agree with the original pattern are incorporated into the novel construction**. If it would not be so, the order would rapidly have dissolved into incoherence. In other words: **order associated with movement appear as the prevailing components of the universal construction**.

# Chapter 9

# The Wizard of Central Park: Stuart Pivar

*"I don't know what his [Pivar's] theory is but there's no question that the development of an egg is not dependent solely on the genes and nucleus, but on the structure of the egg as laid down to some extent. There are proteins that are there. There are non-genetic factors and I wouldn't be surprised if the actual structure of the cell membrane had some influence on the successive divisions that occur."* – **Richard Lewontin**

Stuart Pivar, a chemist and mechanical engineer, lives just off New York's Central Park West in an apartment stuffed with old art and hundreds of natural history specimens – cases of amber-winged and emerald beetles, fuzzy tarantulas, tiny blue scorpions, skulls, skeletons and bottled brains, butterflies, crystals, a stuffed ostrich, extinct Javanese tiger, and a rare topaz once belonging to his friend Andy Warhol.

I met Stuart Pivar 20 years ago when he threw open his place for a fashionable 1980s party. But it was only two years ago that we really first spoke when he emailed me saying he owned the Roman bronze boy statue I was writing about and invited me to come see it.

The statue had been the most expensive item in the controversial sale of Bunker and Herbert Hunt's antiquities collection at Sotheby's, which I'd covered for *The Economist* magazine in June 1990. The bronze boy was going for more than either of the vases signed by the Athenian artist Euphronios. And one of those vases was the Sarpendon wine cup with the same theme as the bowl sold to the Met by Bob

Hecht – the dealer who's been on trial in Rome for conspiracy to loot half of Italy's Greek and Roman culture.

Hecht spread word among the New York dealers (who in some ways resemble the evolution pack) prior to auction that the Roman bronze boy was a fake. And so it went to a Japanese bidder by default. The statue was later returned to Sotheby's where Pivar bought it, had it x-rayed and learned that its joins were ancient. I was fascinated by Pivar's testing of the piece and decided to write a story about the odyssey of the Roman bronze.

While researching that story, I stumbled across Pivar's book on evolution. I was not previously aware Pivar was investigating evolution. Nor was rational morphology a subject I was tracking, so I put the matter aside. But Altenberg rekindled my curiosity about Pivar's experiments.

He gave me a copy of his book *Engines of Evolution* (now *On the Origin of Form*, North Atlantic Books), and told me his work centered on the "dynamic toroidal surface" of the germ plasm as the basic structure of life. He claimed a deformation of the torus could "simulate the full gamut of living forms".

Said Pivar: "The model is like a slow-moving, elongated smoke ring."

In the book, Pivar provides animated blueprints for the human, lobster, fly, leopard, snake and sunflower – all emerging from what he terms the same toroidal germ plasm.

Pivar says it's now apparent that self-organization is where the mechanism of evolution is to be found. In fact, he thinks the quest is over with his torus model.

He is partly inspired by thinkers such as Goethe – who searched for the elusive "Urform" he believed was common to all life – and morphologist Ernst Haeckel.

Haeckel said that the earlier the stage of an embryo, the more it will resemble distant species. But Pivar thinks "the torus is where it stops".

D'Arcy Thompson's pivotal 1917 book *On Growth and Form* is an influence as well. Pivar says he's been working for years in the Thompson tradition, which he characterizes as a mechanistic approach to evolution, using lab simulations, algorithms and topology.

Pivar told me he sent his book to MIT linguist Noam Chomsky when he realized that Chomsky shares the same enthusiasm for rational morphology and that Chomsky wrote back saying he found the concept "fascinating, plausible, publishable". Pivar says he keeps the note on his refrigerator.

Chomsky collaborated some years ago with the late paleontologist Steve Gould – who Pivar considered a close friend – and with cognitive scientist Massimo Piattelli-Palmarini on research into language being based on the laws of structure and growth, rather than on natural selection.

Stuart Kauffman's investigations into self-organization have been important references for Pivar as well.

Through the years Stuart Pivar has been an inventor as well as an entrepreneur – plastics.

Pivar says that at the time he was growing up, his school – Brooklyn Tech – didn't offer courses in biology. He delved into chemistry, physics and math instead. "Ultimately I came up with a mathematical model for evolution," he said.

Pivar has recently made the following statement:

> "Form may indeed have originated, as Stuart Newman says, by nine DPMs (dynamical patterning modules) not by a code in the genes.

> To corroborate his idea I've nominated another DPM, the Primordial Germ Plasm, predecessor of the egg cell. It is in the form of a dynamic toroidal bilayer. See Wikipedia. Too bad the Altenberg 16 have not shown interest in my demonstration that this one DPM acting alone, can simulate the phyletic body plan of plants and animals."

Pivar includes the following scientists among those who see his torus model as "plausible, publishable, and worthy of further investigation": Dimitar Sasselov, Director, Harvard Initiative for the Study of the Origins of Life; theoretical biologist Brian Goodwin (1931–2009); Nobel laureate Murray Gell-Mann; biologist Stan Salthe; NASA origins of life scientist Robert Hazen, and as mentioned, MIT's Noam Chomsky.

# Chapter 10

# Richard Dawkins Renounces Darwinism as Religion

*March 17, 2008*
*9:52 pm NZ*

Atheist evolutionary biologist Richard Dawkins spoke to a packed auditorium at Manhattan's Ethical Culture Society Saturday night about his best-selling book *The God Delusion*. He admitted in a Q&A that followed of being "guilty" of viewing Darwinism as a kind of religion and vowed to "reform" (no one was allowed to tape Dawkins' confession, however, with organizers of the event threatening to march offenders around the corner to the Church of Jesus Christ of Latter-day Saints). I met up with Richard Dawkins the night before at Barnes & Noble in Tribeca where he told me in front of an audience of roughly 200 people (tape recorders were allowed) of the importance of the role of form in making a proper theory of evolution. Dawkins has long been associated with the gene-centered theory of natural selection and is author of the book *The Selfish Gene*. Our Q&A follows:

**Suzan Mazur**: *Richard Dawkins.net* recently picked up my story ["Altenberg! The Woodstock of Evolution?"] about a meeting at Altenberg in July called "Toward an Extended Evolutionary Synthesis," which is believed will move us a bit away from the gene-centered view. Natural selection is under attack and the feeling is that the really interesting evo stuff has to do with form, which we currently have no theory for. I wondered whether you were asked to participate in the Altenberg symposium and what your thoughts are about a remix of the Synthesis?

**Richard Dawkins**: The question is about a recent symposium at Altenberg in Austria.

**Suzan Mazur**: No. It's coming up in July. I was wondering if you were invited?

**Richard Dawkins**: Sorry, it hasn't happened yet are you telling me?

**Suzan Mazur**: No, it's coming up in July, to remix the theory of evolution essentially.

**Richard Dawkins**: About development as well?

**Suzan Mazur**: It seems a move away a bit from the gene-centered view.

**Richard Dawkins**: You've been taken in by the rhetoric.

**Suzan Mazur**: You posted it on your web site – my story.

**Richard Dawkins**: You asked the question: Have I been invited? I'm sorry to say I get invited to lots of things and I literally can't remember whether I was invited to this particular one or not. [some laughter]

**Suzan Mazur**: But it's being viewed as a major event.

**Richard Dawkins**: By whom I wonder. [some more laughter]

**Suzan Mazur**: You might have a look at the story I put up.

**Richard Dawkins**: No. I'm sorry I've got to answer the question now.

I gather that it's an attack on the gene-centered view of evolution and a substitution of the theory of form.

The theory of form I presume dates back to D'Arcy Thompson, who was a distinguished Scottish zoologist who wrote a book called *On Growth and Form* and who purported to be anti-Darwinian. In fact, he never really talked about the real problems that Darwinism solves, which is the problem of adaptation.

Now D'Arcy Thompson and other people who stress the word form emphasize the laws of physics. Physical principles alone as on their own adequate to explain the form of organisms. So for example, D'Arcy Thompson would look at the way a rubber tube would get reshaped when crushed and he would find analogies to that in living organisms.

I see a lot of value in that kind of approach. It is something we can't as biologists afford to neglect. However, it absolutely neglects the question where does the illusion of design come from? Where do animals and plants get this powerful impression that they have been brilliantly designed for a purpose? Where does that come from?

That does not come from the laws of physics on their own. That cannot come from anything that has so far been suggested by anybody other than natural selection. So I don't see any conflict at all between the theory of natural selection – the gene-centered theory of natural selection, I should say – and the theory of form. We need both. We need both. And it is disingenuous to present the one as antagonistic to the other.

# Chapter 11

# Rockefeller University Evolution Symposium

A two-day "Evolution" symposium in May inside Rockefeller University's Buckminster Fuller dome drew a varied crowd of enthusiasts. Athough the event was a gene-centered one – there were some fascinating speakers, such as Roger Buick on the oldest evidence of a specific metabolic process carried out by life on Earth (in a rock in Australia) and Ulrich Technau, a University of Vienna colleague of Gerd Müller, on the Cnidaria and emergence of body features.

As I walked in, I noticed Eugenie Scott in the corner. She's the director of the non-profit National Center for Science Education headquartered in California. Scott was busy typing on her laptop.

I decided to ask her some questions since I'd interviewed her colleague Kevin Padian about the "evolution debate", and he'd hung up on me. Padian was a witness at the 2005 *Kitzmiller v. Dover* trial on intelligent design.

NCSE works with the American Association for the Advancement of Science (AAAS) organizing symposia.

Scott told me she was at the Rockefeller symposium because she was coordinating the lecture that night by University of Chicago biologist Jerry Coyne, who was billed as "the recipient of an Award of Excellence and Meritorious Service from the Illinois Public Defender Association and a John Simon Guggenheim Foundation Fellowship, among other honors".

Coyne's talk was titled: "Feeding and Gloating for More: The Challenge of the New Creationism".

Coyne investigates origin of species from a genetics perspective. He's a pal of *Selfish Gene* author Richard Dawkins. Prior to the symposium, Coyne had asked me not to contact him for future quotes because I told him I didn't need his comment on the Newman & Bhat self-organization paper.

When I introduced myself to Eugenie Scott, who was unfamiliar with my stories on evolution, I asked her what she thought about self-organization and why self-organization was not represented in the books NCSE was promoting?

She responded that people confuse self-organization with intelligent design and that is why NCSE has not been supportive.

I then asked her why she had as an NCSE board director someone from the Church of Jesus Christ of Latter-day Saints-funded Brigham Young University, and suggested that maybe NCSE should reorganize the board.

Scott objected to the comment and returned to her laptop.

At that point I noticed Rockefeller University President, Sir Paul Nurse tip-toe into the dome. I'd been wanting to speak with him as well. I'd emailed some of my evolution stories to him, but received no response.

I finally got the opportunity to chat following lunch, when the elevator in the cafeteria went up to the 8th floor by mistake and Paul Nurse appeared as the door opened. He said he'd never received the stories and that I should try sending them again, which I did by snail mail.

We had another conversation later that day just before the Coyne lecture during which he agreed that a public television roundtable on evolution was a good idea but that Pfizer had sponsored his *Charlie Rose* science series and he no longer had Pfizer as a sponsor.

Several days later, Paul Nurse's assistant called to confirm they'd received my articles in hard copy and she was sure I'd get a call back. When I did not, I phoned requesting an interview. (Our interview follows. Chapter 19 – "Paul Nurse: Revolution in Biology")

Since the Rockefeller conference did not include speakers on self-organization, I took the opportunity to quiz Harvard's Andrew Knoll from the floor following his talk. I asked him if he was aware of Stuart Newman's hypothesis that all 35 animal phyla physically self-organized by the time of the Cambrian explosion a half billion years ago without a genetic program, with natural selection following as a stabilizer. There was a bit of a rustle in the audience.

Knoll had just finished covering life on Precambrian Earth and had taken the opposite, that is, natural selection came first perspective. He said he was not familiar with Newman's paper and insisted: "No, it's natural selection every step of the way." Knoll avoided eye contact with me for the rest of the event.

Washington University Earth scientist Roger Buick told me during the cocktail hour that I'd upset the argument Knoll had just carefully delivered.

But not everyone was upset. I got a tap on the back from Gerry Peretz, the brother of *New Republic's* Marty Peretz who said he was a former student in Stuart Newman's lab. He asked me if I'd like to have lunch.

So we talked. He was careful not to disclose any lab secrets, said he liked Newman and thought he was a superb scientist, although he found his politics a bit too progressive – that Newman had been the darling of *Rolling Stone* at one point.

It may have been a 2004 *Mother Jones* article Peretz was referring to about Newman's attempt to patent a part human, part animal chimera to highlight the dangers of the commercialization and industrialization of organisms, which he fears will ultimately include humans.

The Jerry Coyne address left many speechless – but for the wrong reasons. Why was Coyne preaching about creationism to a highly educated, largely non-religious audience of scientists on Manhattan's Upper East Side? Didn't Coyne know *New York Magazine* ran a "God is Dead" cover decades ago and that churches in Manhattan have turned around in real estate deals for more than 30 years?

Coyne, dressed down in jeans for the talk, and anticipating confrontation, did not take many questions from the floor. So people moved to the stage to engage him before he could exit.

He was not happy to see me. His mouth was white and parched from speaking and he looked like he needed a beer. Nevertheless, he was cordial.

When I questioned his comment in the speech about natural selection (he said he was aware of 300 examples but didn't have time to describe them), and reminded him that even his pal Richard Dawkins said we need a theory of form – Coyne defended his friend, suggesting that Dawkins did not have self-organization in mind...

# Chapter 12

# Mainstream Media Doesn't Get It – Except Vanity Fair

The mainstream media has failed to cover the non-centrality of gene story to any extent. As mentioned in the opening pages of this book, this has to do largely with Darwin-based industry advertising, editors not doing their homework and others just trying to hold on to their jobs. Following is an opinion piece I first submitted to top mainstream news organizations in April 2008, and a sampling of responses. I tailored the story in some cases because of space limitations. Some news organizations got a proposal. Here it is for the record.

**BEYOND DARWINSIM**

> "Unless the discourse around evolution is opened up to scientific perspectives beyond Darwinism, the education of generations to come is at risk of being sacrificed for the benefit of a dying theory." – **Stuart Newman**, New York Medical College

There are distinct parallels to the time of English naturalist Charles Darwin and now. Almost 150 years ago when Darwin set out his theory of evolution in *Origin of Species*, traditional structures were crumbling, particularly in the universities, giving way to the need for more science and technology. Cheap printing brought information to the masses as today's Internet is doing. Secularization swept through Western society and now attempts to complete the job around the globe.

Richard Milner, who tours the US dramatizing Charles Darwin in a one-man show and for many years edited paleontologist Steve Gould at *Natural History* magazine, told me that people were "chafing under the strictures of religion" during Darwin's time. He said Darwin and his disciple Thomas Huxley, nevertheless, were "absolutely amazed" at how quickly the existing paradigm was overturned and Darwinian evolution accepted and adopted not only in science but in law, literature and virtually throughout the culture.

Charles Darwin might also have been shocked to see his image replace that of Charles Dickens on British currency.

From about 1875 on, "science came to be portrayed as a means to create and educate better citizens for state service and stable politics and to ensure the military security and economic efficiency of the nation," British and European intellectual historian Frank Turner has noted. Turner, who I reached by phone at Yale, where he serves as Director, Beinecke Rare Book and Manuscript Library, advised that this is the situation in America today where scientists are "the most successful intellectuals in securing public funds" and in exchange for government grants agree to work to "ensure better health, economic stability and national security".

What does this all mean for evolutionary biology? Well, as civilization experiences another such cycle of "change" – Darwin's theory of evolution has also now come under scrutiny as inadequate in explaining our existence. In fact, there is considerable noise for its reformulation. It was last dusted off 70 years ago and repackaged.

The thinking is we can no longer pretend evolution is just about Darwinian natural selection even if that's what most biologists say it's about and textbooks repeat it.

Darwin didn't even know about genes and DNA, for example. But there is "other" compelling evolutionary evidence emerging, as well, prompting the call for a new synthesis.

In fact, in July a group of 16 biologists and philosophers plan to meet in Altenberg, Austria at the Konrad Lorenz Institute to discuss an Extended Evolutionary Synthesis. Among some of these "other" concepts the KLI group will examine are self-organization and non-centrality of the gene.

One of those participating in the Altenberg meeting is Stuart A. Newman, a professor of cell biology and anatomy at New York Medical College. Newman thinks he has a coherent theory of form – something the Darwinian theory has lacked. It reflects a 20-year synthesis of Newman's work.

Newman's hypothesis has to do with self-organization, which he describes as "the capacity of certain materials, non-living as well as living, to assume preferred forms by virtue of their inherent physical properties". A snowflake is one such non-living example.

In a free-access paper published April in *Physical Biology*: "Dynamical patterning modules: physico-genetic determinants of morphological development and evolution", Newman and his co-author Ramray Bhat conclude that all of today's 35 animal phyla self-organized roughly a half billion years ago as life moved from single cell to multicell (the increased number of single cells at the time of the Cambrian explosion and their secretions caused a sticking together of cells).

A pattern language Newman & Bhat call DPMs (dynamical patterning modules) then explored body building of these highly plastic organisms, *i.e.*, formation of body cavities, segmentation, appendages, primitive hearts and eyes. Selection followed as a stabilizer, say Newman & Bhat.

They also say the DPMs – "gene products and the physics they mobilize" – still exist today, though only to a degree, since they're now "hardwired to the genome through millions of years of stabilizing evolution".

The Newman & Bhat paper challenges the popular view of leading molecular biologist and geneticist Sean Carroll, author of *Endless Forms Most Beautiful*, that an "unusually intense selection on regions of DNA that do not encode proteins led to extremely rapid, but still incremental, diversification of form" around the time of the Precambrian-Cambrian boundary.

Carroll has written in a recent article in *Scientific American* that fewer than 10% of genes "are devoted to the construction and patterning of animal bodies during their development from fertilized egg to adult" and much "remains to be explored".

But Newman, who describes Carroll as a "neo-Neo-Darwinist", emailed me saying that while regulatory mutations can cause changes in form – "sometimes dramatic" – "where do the forms come from in the first place?"

Carroll advised several weeks ago he would review the Newman & Bhat paper, but has now said he's been "hit with a piano over the back – figuratively" and didn't think he could comment. He's also said "biologists are tribal", and that there are several Altenberg-type conferences in the next few months, adding that he declined an invitation to participate in Altenberg.

Richard Dawkins, author of *The Selfish Gene,* told me we need both natural selection and a theory of form during his recent New York book tour. However, he has refused to discuss the Newman & Bhat paper.

Scientists like D'Arcy Wentworth Thompson in the early 1900s pioneered theory of form, although the idea was marginalized. Then in 1972, Niles Eldredge and Steve Gould presented their punctuated equilibrium paper suggesting that evolution has not been a gradual process. And in the past several decades form has been reinvestigated and reinvigorated by developmental biologist Stuart Kauffman's trailblazing of self-organization at Santa Fe Institute, as well as others.

Form appears to be moving center stage as time for discussion of the proposed remix of the theory approaches.

Rutger's philosopher Jerry Fodor's *London Review of Books* article "Why Pigs Don't Have Wings" unleashed the debate about evolution without adaptation last Fall. Fodor is co-authoring a book on this with University of Arizona cognitive scientist Massimo Piattelli-Palmarini and told me the theory of natural selection is "wrong in a way that can't be fixed".

Stan Salthe, a natural philosopher and zoologist at Binghamton University, who hosted a powerful online conversation among scientists about the Fodor article, similarly dismisses natural selection, defining Darwin's theory of evolution as "just unexplainable caprice from top to bottom" and "what evolves is just what happened to happen".

Stuart Newman and Ramray Bhat have schematic drawings to accompany their theory of form. Newman has also written and edited books touching on the subject – one with University of Vienna theoretical biologist Gerd Müller and another with University of Missouri biological physicist Gabor Forgacs.

Meanwhile, Stuart Pivar, an independent scientist in New York, claims he's the only one with an animated set of human and animal blueprints for form – which he's published in his book *Engines of Evolution*.

There is no doubt about the need for a reformulation of the theory of evolution and the public's interest in it. So the fact that this summer's Altenberg meeting, for one, is private has caused some consternation, particularly on the science blogosphere where there's been a call for streaming of the event to a global audience.

Stuart Newman has long maintained the public should share in scientific knowledge because American science is publicly funded. In fact, he's written eloquently about the ethics of this: "In a society with democratic values it should be inarguable that those who pay for scientific research and will eventually experience its effects should be informed of what is in store while there is still a chance to discuss its objectives and influence its course."

Don't we all really want to know how publicly funded scientists are rethinking our evolution?

**Publication 1:**

> "Dear Suzan:
>
> I am writing to you from [redacted], where I am on vacation.
>
> I made a joke in passing to you a couple of weeks ago, a sour little joke about the nature of journalism, but little did I know the whole thing would come almost completely true.
>
> I must explain.
>
> As you know, I had to run a piece by [redacted]. Her thesis was that there is a paradigm shift going on. I thought it was a great piece.

My joke to you was, "So let's get [redacted]'s piece in the paper – and then we can run yours, unless my new bosses hate [redacted]'s piece and it gets me fired."

Well, it didn't get me fired, not yet, anyway, but boy, did they hate it. Too complicated, they said. Not clearly enough explained. Ahead of the audience. Gotta be careful, they said; as editors, we have to come out of ourselves and think of our readers more.

...in fact, said they, we should begin to rethink the whole... section, which (despite recent readership numbers which were record-setting) is probably too brainy and needs more edge, more appeal to young folks, a re-design. And they told me point-blank not to try anything like [redacted] again.

Whew. This means we will not run your piece... I... apologize, very sincerely, especially in light of your patience and kind willingness to add things I was wanting.

... I am grateful to you for your hard work and devastated about this turn of events. Sincerely"

**Publication 2:**

4/24/2008

"Suzan,

Thanks very much, but I think we'll pass. This isn't quite right for us. Also, we currently have a staff writer and two contributors covering evolution so the field is fairly crowded for us. Best, Jim Gorman [*New York Times*]"

**Publication 3:**

4/23/2008

"Suzan

This has a lot in it that's interesting, but I think that it's just going to take more space than we have on offer to adequately explain what these scientists are up to for a general readership. We're going to pass on it, but thanks very much for letting us consider it. I hope you'll keep us in mind in the future. Best, Susan [Brenneman, *Los Angeles Times*]"

**Publication 4:**

4/28/2008

"Hi Suzan,

Sorry to be so slow in getting back to you, I am going to pass on this. The evolution issue is an interesting one and I'm sure you'll be able to land this somewhere. Best, John Pomfret, Editor, Outllook, *Washington Post*"

**Publication 5:**

6/3/2008

"Suzan, while your argument undoubtedly deserves an airing, I still think our mag is not the right venue for this piece. We don't usually stray into scientific debates in this kind of way. I urge you to try somewhere else–if not a popular science journal, then maybe one that's less focused on politics/culture than *The Nation*. Regards, Roane [Carey]"

**Publication 6:**

6/5/2008

"Dear Susan,

Thanks for your call, and for this pitch. It's a good idea for a story but, as I should have mentioned yesterday, we've actually got a Darwin piece in the works already. It's much smaller in ambition than yours, but I'm afraid it crowds out the possibility of our assigning another one. So we'll have to pass. But I wish you luck in placing this elsewhere. Best, Scott [Stossel, *The Atlantic*]"

**Publication 7:**

2/13/2008

"Arghhh! Sorry I didn't phone. Have only just finished my section, and frankly the only conversation I want now is with a bottle of wine.

Did Fodor say anything interesting?

BTW, I see from your code that you are a New Yorker. I'll be there at the end of next week. Given that I wouldn't be able to run your story in next week's edition anyway, as SciTech will be devoted to a conference report from the AAAS (which is why I am coming to America in the first place), maybe we could thrash things out face to face over a different bottle of wine then? Bestest, Geoff [Carr, *The Economist*]"

**Publication 7 (more):**

2/22/2008

"Sorry. Have had to cancel the New York trip. Will be in touch when I get back to London. Geoff"

**Publication 7 (more):**

3/20/2008

"Dear Suzan I'll be in town next week. What does your dance card look like? Best Geoff"

**Publication 7 (more)**:

3/24/2008

"How does Wednesday morning look? Geoff"

**Publication 7 (more):**

3/25/2008 11:54 AM

"I'm pretty much chocca from Wednesday evening to Friday evening, I'm afraid. I've been roped into a conference. Coffee on Saturday morning? Or I'll be back again in about three weeks.

What are we discussing? I think we are discussing the difference between evolution and ontogeny, and whether there is any fundamental mechanism of evolutionary change other than natural selection. . . Geoff"

**Publication 7 (more):**

3/25/2008 2:38 PM

"Well, I'm still very sceptical of the whole premise. You would have to convince me it is intellectually sound before I commissioned something. There is some interesting stuff. Most of what is being presented as new either still depends at bottom on natural selection on the primary DNA sequence or does not seem to me to have the necessary generational stability to act as an alternative mechanism. Do you reckon you could do that? Geoff"

***Vanity Fair:***

6/16/2008

"Hello Susan,

We've reviewed your pitch, and we'd like you to follow up with us after the Altenberg conference to let us know what came out of it, and how the scientific community is reacting. At that point, we'll take another look and get back to you. Best Wishes"

# Chapter 13

# Stuart Newman: Evolution Politics

*August 24, 2008*
*11:00 pm NZ*

> *"If Newman is right, his theory explains why all animals have an increasing number of bones from their shoulders to their digits. As they sculpt the human form, embryonic cells almost seem to be talking to each other."* – ***Newsweek***, "How Life Begins," January 11, 1982

It's not surprising that Stuart Newman was one of "the Altenberg 16" scientists who kicked off a reformulation of the theory of evolution, the Extended Evolutionary Synthesis, this July at Konrad Lorenz Insitute. While I've been writing about Newman's work over the last several months, his scientific investigation into form (limb bud development) was first showcased to an international audience a quarter century ago – in a 1982 *Newsweek* cover story on the embryo. Since then Newman, a dedicated cell biologist and professor of anatomy at New York Medical College, has been in and out of the news, writing about the ethical issues of human genetics and bioengineering in scientific journals, sometimes appearing on public television, as well as testifying before Congress when asked.

Stuart Newman's current hypothesis is that all 35 or so animal phyla physically self-organized by the time of the Cambrian explosion a half billion years ago using what he and his co-author Ramray Bhat call a pattern language – DPMs (dynamical pattering modules).

The DPM concept has generated excitement since publication in *Physical Biology* in April, although the commercial media is just beginning to notice.

Newman is a patient man, doesn't take being overlooked personally, and attributes the lack of mainstream media coverage to "disseminators of information" not yet understanding a physical approach to evolution. Newman, on the other hand, has an A.B. in chemistry and a Ph.D. in chemical physics and is largely self-taught in biology.

With the highly publicized Altenberg meeting over, Newman felt comfortable enough to suggest that I visit him at New York Medical College in Valhalla "one day," where he teaches and directs a research lab, to talk more about his work.

He followed up with this polite email:

> "Please take the Metro-North 1:48 pm "Southeast" train from GCS [Grand Central Station] to Hawthorne, NY. I'll pick you up at the station at 2:30 pm (just come down the stairs)."
>
> (Just come down the stairs)? I wonder why the parentheses...

Several days later, I take the train to Hawthorne and walk down the stairs to the parking lot to meet Stuart Newman. Am I misreading cues or has he just spotted me through the window of the enclosed stairwell, put his hands in his pockets and turned his back?

He later asks, "Did I really do that?"

Stuart Newman is a graceful man, about 6'1" with the hands of a microsurgeon – which he is. He is dressed in casual European elegance with sleeves turned up. I try not to be, but am affected by his sincerity and focus. There is an exotic twist to his hair, which in earlier photos makes him look North African.

Newman keeps fit on a vegetarian diet and does not follow sports, although he confesses he's been watching the Beijing Olympics. He says he's never attended a New York Giants football practice though, despite the fact that the Giants summer camp at Pleasantville for many years was only a village or two away from his office. A former student describes Newman as "a cerebral guy".

We drive to Newman's lab in Valhalla, a leafy village with fewer than 8,000 people situated along the Hudson River about a half hour from Manhattan.

I finally meet his charming grad student Ramray Bhat, who I've also interviewed. Bhat is from India and we speak briefly of problems in Kashmir, a conflict I covered in the 1990s.

The first thing I notice as I follow Stuart Newman into his office is a collection of champagne bottles against the far wall. Newman says they represent the thesis defenses of his graduate students. One bottle of Mumm Extra Dry is from March 1970 in celebration of Newman's own Ph.D. defense.

Newman offers me a chair beside him. His computer opens to his desktop screen containing the voluptuous images of Rubens' "The Apotheosis of Henry IV and the Proclamation of the Regency of Marie de Medicis on May 14, 1610". He says art was his first love – he still draws – and then science.

He next gives me a crash course in self-organization, presenting the visual evidence:

> "This is an extinct limb. And this is another extinct limb. This is a modern limb of an amphibian. This is a bird. An iguana. And you can see that they're all kind of built the same way. They all have a single bone and then two bones. And maybe a cluster of bones. There's a mathematical regularity.

This is a transparent view of the chicken limb as the bones start to emerge. And the ones close to the body differentiate first before the ones furthest from the body. It's the same for all vertebrates – definitely all birds and mammals. They're showing the orientation. It's called proximal - distal, dorsal - ventral, anterior - posterior. They're just axes.

Here's a limb bud. It's confining itself to the tip where fingers form by cells contacting each other which then turn into cartilage. Some of the cells in between the dotted lines die off. They don't interact, they just die off. Or in a duck's foot they become webs. They don't differentiate into cartilage.

What happens when these cells interact is that they undergo a process of condensation. There's a clustering. This actually becomes one of our DPMs – the ability of cells to respond to their microenvironment and cluster…

I'll show you what a self-organizational process looks like. So here [looking at cells clustering] are places where it starts up randomly, some then fade away and some get stronger. With self-organization, you can have random starts at different places but then you have competition between the centers and finally you get a pattern, which is going to oscillate. The pattern is going to subside and then it's going to come back. And it will come back with the same statistics but the peaks will be in different places. That shows it is a true self-organizational process…

> We've taken this self-organizational idea and put it into the context of the geometry of the limb. And we've said that at the tip of the limb there's something suppressing it from happening. Cells have to escape from this suppression to organize into spots or rods.
>
> The geometry changes subtly as the limb grows in length. Under some conditions you'll get one skeletal element. Under other conditions you'll get two. Under still other conditions you'll get three or, as in the human hand, five."

Newman closes out the program and shows me his previous screen, a much more ethereal image. I wonder what the Rubens says about who Stuart Newman is now...

On the way out of the office we pass through his lab where he opens an incubator tray of fertilized eggs that his students are observing.

It begins to rain as we exit the college and head into Tarrytown for tea. We park not far from the Tarrytown theater where the Jefferson Starship will soon appear.

Tarrytown's Silver Tips is one of the most "serious" tea rooms in the tri-state area, offering 140 kinds of tea. We settle in at a table and order a pot of Assam, which comes in English Chatsford china with matching cups. The feel of the Chatsford cup is half of the delicious experience of sipping. Talking with Stuart Newman naturally is the other half.

Newman mentions his postdoc days at the University of Sussex and his fondness for English scones with clotted cream and preserves, which he now has a chance to enjoy again.

"You don't find clotted cream around much," he says with a certain nostalgia, as he dips into the cream and raspberries.

How does the Tarrytown scone compare with the Sussex scone?

It's "authentic" but "too much for me actually... won't you have some?" he asks.

He describes his high school years (same one paleontologist Steve Gould attended) and tells me a bit of his family history. We kibbitz about the Catskills and evolutionary politics.

The rain lets up as we leave the tea room and walk downhill to the car. Newman walks in front of me and begins to pick up his pace telling me he forgot to put money in the parking meter. Luckily the car's still there.

Prior to my visit, Newman sent me an email asking if I'd like to see Usonia, a colony of homes built in the 1950s by Frank Lloyd Wright and others in the woods of Pleasantville. I was fascinated by the idea.

So we drive to Pleasantville along the Kensico Reservoir and then onto Route 141. Newman is a bit concerned that we're losing sunlight. We turn right now on Lake Street, right again on Bear Ridge Road and make another right onto Usonia Road. Fifty Usonian homes made of glass, wood and stone are somewhere in the surrounding hills, three of them designed by Wright. And we are about to try to find some.

Usonia was begun by a group of New Yorkers following World War II who pooled $22,000 to purchase 95 acres in the area, eventually creating their own homes at a cost of about $5,000 each. Today the homes are individually owned (but the community spirit survives) with some original residents still living in them.

It's about 6 pm as we enter the woods. Interesting shadows are at play. Newman's velvety voice becomes even more so as he whispers, "It's like we're stalking wild animals."

"That one is a Wright house, isn't that something!" He points to the home with a section covered with field stones and amber light oozing from the windows into the trees. "You've got to come back in the winter and we'll…"

Newman says he was never crazy about Wright's design of the Guggenheim Museum though, describing it as "an affront to its surroundings". "Wright's concept was that everything was supposed to conform to the setting and then he plunks this thing on Fifth Avenue which has nothing to do with Fifth Avenue. I think he just didn't like New York City."

You're only allowed to drive through Usonia, and are not supposed to leave the road, but Newman says he often winds up in somebody's back yard. He points out a sculpture garden. And a tennis court.

Mel Smilow, a famous furniture maker used to live at Usonia. Newman says Smilow and his wife were involved in the nuclear freeze movement and that he got to know them and their house then.

But the shadows soon grow longer, so we leave the enchantment of the forest and head for the Hawthorne train back to Manhattan.

Newman waits with me on the platform as four or five trains pass all going in the wrong direction. 7:21 pm comes and goes without a New York bound train. 7:47 and still no train. We soon learn that trains to GCS are off-schedule because a tree fell onto the track several stations away and there is no announcement about when service will resume.

I have to persuade him that I am a veteran of several wars before he agrees to leave me – insisting that I call him at home if there are further delays (and there is an email later from him asking me to email him as soon as I return home).

We say goodbye. And Stuart Newman disappears into the night and nearly full moon.

Stuart Newman is co-author of the textbook *Biological Physics of the Developing Embryo* (Cambridge University Press) with Gabor Forgacs, and with Gerd Müller (Chair, Konrad Lorenz Institute) co-edited *Origination of Organismal Form: Beyond the Gene in Developmental and Evolutionary Biology* (MIT Press), a volume about the origination of body form during Ediacaran and early Cambrian periods, also contributing a few chapters to it

Newman's A.B. degree is from Columbia University and his Ph.D. in chemical physics from the University of Chicago where he also did post doctoral studies in theoretical biology, as well as at the School of Biological Sciences, University of Sussex. Newman's been a visiting professor at Pasteur Institute, Paris; Commissariat a l'Energie Atomique-Saclay and the Indian Institute of Science, Bangalore.

The text of our recent Valhalla interview follows.

**Suzan Mazur**: Stuart, you are one of the 16 scientists who met this July in Altenberg at Konrad Lorenz Institute to talk about reformulating the theory of evolution based on natural selection. Is the Extended Evolutionary Synthesis now a reality following the Altenberg meeting?

**Stuart Newman**: Yes, I would say it's a reality not just following the Altenberg meeting but people who were there and others who weren't there have been working on extending the breadth of the evolutionary synthesis for some time from a number of perspectives. This meeting was just a confirmation that it's underway and perhaps an impetus to push it forward even more.

**Suzan Mazur**: The extended synthesis has been described as a graft onto the modern synthesis, the Neo-Darwinian theory of evolution. While your final report is to be published next year by MIT, is the plan in the interim to cast a net globally for scientific perspectives on the extended synthesis, because as it stands now the extended synthesis is largely an American-European concept?

**Stuart Newman**: Yes. I know people in India and Japan who are working very much along these lines and extending the synthesis. We're all in touch with our international colleagues and I don't think it really is confined to the US and Europe.

**Suzan Mazur**: Is the modern synthesis, the Neo-Darwinian theory of evolution, now history so that public money being spent on research based on the modern synthesis may in fact be wasted?

**Stuart Newman**: I wouldn't put it that way. Up until fairly recently what's called the modern synthesis has been the only game in town. I don't think that anybody at our meeting or any people working in evolutionary theory – maybe there would be some that would say that the phenomena and mechanisms described in the modern synthesis don't pertain anymore. They do pertain. The question is: Are they the major forces of evolution? Are they somewhat subordinate?

There is a wide spectrum of opinion even within our group of 16 as to whether the mechanisms of the modern synthesis are the predominant mechanisms or subordinate mechanisms.

**Suzan Mazur**: How much agreement is there among your Altenberg colleagues about the gene arriving late in the evolutionary process and playing a secondary role?

**Stuart Newman**: I think possibly some things that I've said, or possibly some other people have put in that light, suggest that the gene is a late arrival – but we never in our scientific papers have said anything like that. We're all dealing with organisms that have genes.

Even before there were multicellular organisms, there were single-celled organisms with several billion years of evolution behind them in which genetic mechanisms were refined and established. Before that there's pre-biotic evolution that led to single cells that had genes. But once we reached the multicellular world, which is maybe half a billion years ago or more – but not a very long time relative to the entire history of cellular life – once we hit that stage, we're very much dealing with organisms with genes.

The question is: All of the complex forms that have emerged since that point, since multicellularity, are they dependent solely on genetic mechanisms or is it genetic mechanisms plus other determinants of morphogenesis for the generation of forms? I don't know anybody who thinks genes arrived late. But perhaps genetic consolidations of patterns and forms trail after the origination of the patterns and forms.

**Suzan Mazur**: With the extended synthesis, has the shift actually begun from the gene-centered perspective of evolution to non-centrality of the gene? As a result of this meeting [Altenberg], has the shift happened?

**Stuart Newman**: In some people's minds the shift happened a while ago in their own work. The general shift has not happened, and it may not happen for quite some time, if it happens at all. But let me put it in a more precise way. It's not just a shift from mechanisms that use genes to mechanisms that don't use genes. All the mechanisms use genes.

The question is are there different forms solely due to genetic evolution or are they due to other organizing processes of multicellular life beyond the gene – in addition to the gene?

**Suzan Mazur**: What do you think the origin of the gene is?

**Stuart Newman**: The origin of the gene is one area that is very poorly understood in evolutionary biology. Here we're talking about what's called pre-biotic evolution – evolution that preceded cells as we know it, because all cells that we know use genes as their hereditary mechanism.

There was a period in early cellular life – and this has been written about by a scientist named Carl Woese. He believes that genes were exchanged very freely between organisms. He calls this a pre Darwinian world where basically organisms take on new genetic characteristics not through what's called vertical transmission but by horizontal transmission – from sharing with other organisms.

If you go back even further – where did the genes come from? Some people suggest there was a world of RNA that preceded the world of DNA and that RNA was the most primitive genetic mechanism. So we might have RNA genes preceding DNA genes. But where did the RNA genes come from?

Then you get into the area that's called chemical evolution. There are processes in non-living systems that could potentially generate the molecules of living systems. There are theories of thermal vents under the ocean where there's a very extensive chemical shuffling and reactions and things that could potentially generate these building blocks.

I've been to meetings where this has been discussed and I've met some very smart people who are at work in this area, but there are still big gaps in our understanding.

**Suzan Mazur**: The term "punctuated evolution" is part of the new extended synthesis language. How does it differ from the Eldredge-Gould concept of "punctuated equilibrium" – which argues that evolution has not been a gradual process, as reflected in the fossil record?

**Stuart Newman**: This is a somewhat controversial area because it involves, first of all, estimations of time scales. When Gould & Eldredge first presented their idea of punctuated equilibrium, they were just pointing to the fossil record and saying that things that were considered gaps in the fossil record because there were discontinuities between the forms that were uncovered, might not truly be gaps. Things may have happened relatively fast. And therefore, one form supplanted another form by evolutionary processes that were not entirely gradual.

Over the years I would say there was a backing off by Gould and Eldredge from considering what evolutionary biologists would call saltational mechanisms, which are true jumps. So you have one form being replaced by another form very rapidly. Gould & Eldredge would both talk about rapid in a geological sense. They were talking about millions of years of evolution rather than hundreds of millions of years of evolution.

From our own work and the work of other scientists looking at physical determination of biological form, there's a very good case to be made – and there's a lot of experimental data behind it – that true jumps are possible.

To give you an example, we now know the mechanism by which the number of segments of organisms like ourselves are generated. Our backbones have serially repeated vertebrae. There are a certain number of them – 30, 40. And some snakes have 300.

We know that the process that generates these blocks of tissue involves an oscillation, a clock-like mechanism. This kind of beats time, and the time is translated into spatial discontinuity. Whenever you reach a certain point on the clock, you've cut off another block of tissue. And you do it again.

That clock can run faster or slower. What makes the clock run fast or slow is – even if it's a small change, can lead to a transition from 40 blocks of tissue to 300 blocks of tissue all in one step. So you can get saltational jumps. It doesn't have to be gradual. You don't have to go from 40 to 41 to 42 to 43 up to 300. You can get much bigger jumps than that. We know that is the case.

Even though Gould & Eldredge backed off of saltational mechanisms, we now know from embryology that saltational mechanisms are very plausible mechanisms of evolutionary change.

**Suzan Mazur**: Your theory of form also describes an evolutionary spontaneity. You say that all 35 or so animal phyla physically self-organized by the time of the Cambrian explosion half a billion years ago using a pattern language – dynamical patterning modules (DPMs) – and that selection followed as a "stabilizer". Is that correct?

**Stuart Newman**: Yes.

**Suzan Mazur**: As these terms apply to your hypothesis, what do you mean by "self-organization" and "selection"? Do you mean selection in the Darwinian "survival of the fittest" sense?

**Stuart Newman**: Yes – to some extent, and there are other ways selection acts as well. First of all, by self-organization we mean when cells get together and form clusters there are physical processes that are relevant to material on that scale. A single cell is a very small object and it's subject to certain physical forces. Cells can be knocked around randomly if they are in a fluid medium.

When you get to larger structures, you have things operating like diffusion, the flowing of materials from one end to the other that can cause non-uniform gradients. And you have cells in clusters, some of them being more strongly adhesive and some less strongly adhesive. So you'll get phase separation – like two immiscible liquids, oil and water. You can get different layers of tissue due to this process.

Then you can have interactions between cells where a cell will exert some inhibitory effect on the cell next to it. Cells right next to one cell won't do the same thing. These are all processes that use molecules and genetic means that were evolved for single cells but in the new multicellular context along with the physical processes that are characteristic of larger scale matter. Again, organizational principles kicking in that just weren't there in the single cell state.

Cells have these clocks inside of them, these oscillations. And in the single cell world an oscillation just periodically changes the state of a cell. But in the multicellular state, the oscillation can lead to spatial segmentation. You're mobilizing things that existed before, that evolved in the single cell world but then when they meet up with the physics of mesoscale (middle scale) materials, you get all these morphogenetic processes – all these form-producing processes come into play.

To give an example, a molecule of water doesn't have waves and it doesn't have whirlpools. It's just a molecule and it has the physics of molecules. When you have a lot of molecules of water, they make liquid water. Then you can get all sorts of disturbances and wave-like and vortex-like phenomena, which you would never see in the individual molecule.

It's not that the molecules have changed but the scale has changed. Some new physics comes in and organizes the system.

**Suzan Mazur**: The selection you're referring to though is not Darwinian survival of the fittest…

**Stuart Newman**: Let me go on from there. The forms that you get are not due to Darwinian selection. They're due to the inherent properties of the system. But many of those forms may not be viable. You might get forms due to physical organization that are not suited to this world. They just can't find a place to live. They can't eat or whatever. There will be a shake-out phenomenon. Some of them will survive and others won't.

In that sense the Darwinian mechanism of selection is a kind of culling process. It doesn't create the forms but it basically determines which ones of them will persist in the world. It is a role for Darwinian selection but it's not a role of building up forms in an incremental fashion. Or at least by and large it isn't. It's culling these self-organized forms and just selecting among them.

**Suzan Mazur**: Many scientists say the terms "self-organization" and "self-assembly" are interchangeable. Some say the snowflake forms by self-organization – you call the snowflake formation process self-assembly.

Some scientists describe the self-assembly of the hydra, for instance. You did extensive experiments with the hydra in the early 1970s in England at the University of Sussex, and term the hydra's regenerating process self-organization. To what do you attribute the difference in use of these terms self-organization and self-assembly? And what do you mean by "self-assembly"?

**Stuart Newman**: Self-assembly is an area that comes out of chemistry and physical chemistry. Molecules can undergo a process of self-assembly. It depends on the shapes of the molecules. Very often you see self-assembly where you have a number of identical or nearly identical sub-units and they come together and they form structures that have certain variable properties to them and certain common properties.

Because of the molecular nature of water, it will crystallize with certain symmetries and when it forms snowflakes they'll always have six-fold symmetry but every one of them will be different from the other. This is the self-assembly process. It's not a process where the form depends on a flux of matter and energy. You have subunits coming together. It's generally considered an equilibrium process. Some self-assembly processes utilize some energy but by and large self-assembly is reserved for the equilibrium processes. And you get static structures.

A snowflake, as long as you keep it cold, will stay there. It doesn't require a constant input of molecules and an expulsion of molecules. It's a static structure.

Self-organized processes often require a flux of matter or energy just to keep the structure in place. The physical chemist Ilya Prigogine, a Nobel laureate who looked into the self-organization of matter, called these things "dissipative structures". They're structures that remain in place by using energy to stay in form. Many of the self-organizing processes are like that. They're dissipative structures.

When I talked about this process of generating the vertebral column – it uses chemical oscillations. It uses oscillations of molecules inside the cell. You can't have an oscillation without expending energy and having a flux of material in and out of the system. In that sense it's a non-equilibrium system, but eventually it generates static forms if the forms get "frozen" into place by cell differentiation.

So there's a big distinction from the point of view of physics and chemistry between self-assembly and self-organization. I want to hasten to say that inside the cell there are processes of self-assembly. Many of the cytoskeletal structures – the structures that keep the cell in shape internally – are made up of proteins that self-assemble: microtubules, microfilaments. You have an identical collection of a small number of subunits and you have many copies of them, and they will form very long extended structures.

**But most of those molecules had to evolve to be able to self-assemble**. You can't just take any old molecule and expect it to self-assemble. So self-assembly that you see inside the cell is a function of subunits that, in general, have evolved to self-assemble.

**Self-assembly *per se* can't explain evolution**. It happens in the physical world but when it happens in the biological world, the units that self-assemble are products of evolution.

**Suzan Mazur**: Is it because self-organization is so misunderstood within the biology community that you've tucked it under the umbrella term "phenotypic plasticity" in the extended synthesis, along with epigenetics – plasticity being a concept generally understandable?

**Stuart Newman**: Yes. We had a summary from the Altenberg conference. We wanted to create a number of umbrella categories under which all the things we discussed could be subsumed. The biologically interesting aspects. Self-organization is a concept imported from physics and physical chemistry.

Its biological implication is that if you have an organism with a certain genotype, then because of its self-organizing processes – which are subject to perturbation from the environment, etc. – you can get different outcomes with the same genotype. From the point of view of evolutionary theory this is plasticity, the ability to make different forms from the same genome.

It wasn't important to enforce every mechanism of plasticity. Plasticity was a sufficient umbrella to include those concepts.

Plasticity is not only associated with self-organization. Molecular self-assembly can also be plastic. It is now recognized that many proteins have no intrinsic three-dimensional structure – their forms and functions change depending on their microenvironment, including other proteins that may or may not be present. The structure and function of macromolecular complexes can therefore change dramatically over the course of evolution with minimal genetic change, or as a side-effect of other changes, not driven by adaptation. This is quite relevant to the evolution of highly complex structures like the bacterial flagellum, a problem constantly harped on by advocates of intelligent design.

One reason that so little progress has been made in this area is that perfectly valid scientific concepts that employ nonadaptive evolutionary mechanisms are rarely considered because of the hegemony of the neo-Darwinian framework.

**Suzan Mazur**: Why do you suppose the mainstream media stonewalls coverage of self-organization when the National Science Foundation supports the investigation, including your own? The research is there. Self-organization is real – so why the media gag?

**Stuart Newman**: I don't know if it's a purposeful gag. I think in order to understand self-organization and not just the term, it really takes a certain level of sophistication in the physical sciences as well as the biological sciences. People can understand what a gene is and how a gene specifies a protein. Those ideas are easy to explain to the public. Ideas of self-organization are much more difficult to explain. I feel that many people in charge of disseminating information don't understand the concept. Just the way many physicists don't understand biological concepts, many biologists don't understand these physical concepts. There's a two (or more) culture problem in theories that are relevant to evolutionary biology. It's a matter of people not understanding the concepts well enough.

**Suzan Mazur**: To what do you attribute the reluctance to distribute literature about self-organization by organizations like the National Center for Science Education?

**Stuart Newman**: I think there is a challenge that self-organization and plasticity in general presents to Darwinian theory because Darwinian theory is basically a theory of incremental change. If you are confronted with a very complex structure, the Darwinian explanation is that it was built up little by little over long periods of time. If you bring in self-organization, then that undermines those Darwinian explanations.

To my mind, self-organization does represent a challenge to the Darwinian, *i.e.*, the modern synthesis and the perceived understanding of evolutionary theory. People are concerned – though I don't agree with them for being concerned about it – but people are concerned that if they open up the door to non-Darwinian mechanisms, then they're going to allow the creationists to slip through the door as well.

This is not at all the case. But because many of evolution's designated defenders don't understand these concepts well enough, they are fearful of that potential.

**Suzan Mazur**: The National Center for Science Education director Eugenie Scott told me that her organization does not support self-organization because it is confused with intelligent design, *i.e.*, "design-beyond-laws" – as Michael Behe, a biochemist at Lehigh University describes it. NCSE also pays lucrative fees to conference speakers who keep the lid on self-organization by beating the drum for Darwinian natural selection. NCSE and its cronies completely demonize the intelligent design community, even those who agree evolution happened. Religion is not the target since even the National Academy of Sciences embraces religion. So it seems the real target is those who fail to kneel before the Darwinian theory of natural selection and prevent the further fattening of the Darwinian industry tapeworm.

NAS and NASA/NAI in their respective publications *Science, Evolution and Creationism,* and *Astrobiology Primer* have also kept out any discussion of self-organization. What is your response to this? Why do you think such organizations continue to feed unenlightened information to the public at public expense?

**Stuart Newman**: Although I may not use all the terms that you used, I would have to agree with you that if you look at the Pennsylvania legal case on the teaching of evolution [*Kitzmiller v. Dover*], there was a very solid identification of evolution with Darwin's theory of evolution. I think that this was very reinforced in the public mind – that if you believe in evolution, you believe in Darwin's theory of evolution because it's supposedly the same thing. And if you don't believe in Darwin's theory, you must believe in something supernatural.

**This is not at all valid and I think it's a big mistake because we know there are non-linear and what I call saltational mechanisms of embryonic development that could have contributed – and I'm virtually certain that they did – to evolution. It was Darwin who said that if any organ is shown to have formed not by small increments but by jumps, his theory would therefore be wrong.**

The people you refer to – instead of moving beyond and expanding Darwin's ideas to include things like self-organization and bring other mechanisms into it – adhere to this Darwinian orthodoxy where everything has to be incremental. And when confronted with something very complex like the bacterial flagellum or the segmented vertebral column, they say that it had to have arisen in an incremental fashion.

But there are other mechanisms involving self-assembly and involving self-organization that could potentially explain these things as long as one did not seek purely incremental explanations. And physics and the theories of self-organization show us that those mechanisms exist. I think it's an unfortunate error that some advocates of evolution are making by adhering so closely to this incrementalist Darwinian dogma.

**Suzan Mazur**: What is the danger in mediocre science being pushed on the public, aside from the wasting of public funds at a time of serious economic downturn in America?

**Stuart Newman**: It seems to me that if somebody is predisposed to be skeptical, perhaps because they are religious, and are told that the vertebrate column, for example, had to have evolved incrementally – they may not be persuaded by it because it's not true, even though their motivation not to be persuaded might come from their religion. Then scientists who are working on this embryonic mechanism who have shown that there are non-incremental mechanisms that produce these things come along, and therefore everybody who's been assuring this skeptic that it's all incremental turns out to be wrong.

It really undermines confidence in science if people are always being subjected to what we call handwaving arguments that all complexity had to have had an incremental origin.

**Suzan Mazur**: Sam Smith of *Progressive Review* recently said the following: "[Scientists] are also subject to that most pernicious of academic temptations: the desires and biases of their funders." He refers to the "distorting role of the Defense Department, agribusiness and pharmaceutical corporations in supposedly objective science." Would you comment on that?

**Stuart Newman**: That's true. I don't know how pertinent it is to the evolution debate. I don't think the pharmaceutical companies have a role in steering the field away from self-organization. In fact, if it's true and you can patent it and make money on it, they'll chase after that. So I don't think the businesses, although they do have this grip on scientific development – I don't believe they're ideological in that way. The ideology really comes from entrenched old science –

people who are educated in biology with no sense of physical sciences. The inertia and obduracy comes from the side of the scientific establishment rather than from industry.

**Suzan Mazur**: As a common language has begun to be created by you and your Altenberg colleagues in an effort to help bring various scientific fields together – will the public's need to know be taken into consideration as well? Do you envision a kind of "science-ese" happening, a plain speak on the order of "legal-ese" – the attempt by the legal community to simplify document language because law firms have gone global?

**Stuart Newman**: There are people working on all sides of science who want to communicate with the public. The loudest voices and sometimes the more facile and appealing writers have been retailing old ideas unfortunately. As younger people come into the field who are more open-minded, some of them will also have a facility with public communication and newer ideas will get out. People will understand that evolution happens, but that it doesn't necessarily happen the way Darwin said it happened. Sometimes it does. But it happens other ways as well.

It will be to the benefit of everybody if evolution is acknowledged and incorporated into the broader culture, which it isn't now. It's kind of in a losing battle. The only way that can happen is by not communication of bad ideas but by communication of good ideas.

**Suzan Mazur**: But can it be communicated in a simpler way? Can the jargon be broken down so the public can get a handle on it?

**Stuart Newman**: You can use metaphoric means of communication. I tried to do it talking about water having waves and having vortices and it being something different from the molecules that the water is made of. I think people are more responsive to that approach than to assertions that every complex organism or structure had to be built up by incremental means. People don't buy that because it's often not true.

**Suzan Mazur**: How much of a makeover of the evolutionary biology community will the extended synthesis entail?

**Stuart Newman**: If you polled different people at the Altenberg meeting, some would say well it just needs to be what already exists plus a few additional things that they are working on. But I think the bigger challenge, particularly coming out of this idea of plasticity, is to the idea that things are built up in increments and that it's only genetic change that drives evolution. I think that idea is a big idea. And that big idea is coming up against a lot of entrenched belief within the scientific community that things happen in the Darwinian fashion.

**So I think ultimately it's going to be a big turnaround in evolutionary theory even though it might look like it's happening slowly.**

**Suzan Mazur**: Massimo Pigliucci, one of the coordinators of the Altenberg event, has told me the extended synthesis will not affect the lives of people in general. Obviously you don't agree?

**Stuart Newman**: I think it will affect the lives of people in the sense that right now we have these walls around belief in evolution or non-belief in evolution. We've moved past having big wars about whether people should take certain medications for high blood pressure. You could be religious and take blood pressure medicines. Or you can be an atheist, etc. We don't have wars about many aspects of science.

We do have wars about some of the aspects of science. And one of the reasons we have these wars is because people are being asked to accept implausible and incorrect mechanisms.

If better ideas of how evolution occurred get out into the wider culture, there will be more of an acceptance of the phenomenon of evolution. People will stop fighting about whether evolution happened, which is a ridiculous fight. And it's partly a ridiculous fight because of religion. The other side is because of science. Because the science is not where it should be. For the video interview, see: http://video.google.com.au/videoplay?docid=3516772316650379357&hl=en#

# Chapter 14

# The Astrobiologists

## Bob Hazen: The Trumpeter of Astrobiology

*July 30, 2008*
*4:02 pm NZ*

After reviewing Robert M. Hazen's 28-page C.V. of his work as an experimental mineralogist (listing grant $$$ too), as an educator, author of a dozen books and a symphonic trumpeter – I was most charmed to read that a new mineral "precipitated" by microbes in California's highly alkaline Lake Mono had been named in his honor in March of this year: "hazenite". Bob Hazen is the Clarence J. Robinson Professor of Earth Sciences at George Mason University. For the last 30 years, Hazen has also been a scientist at the Carnegie Institution of Washington's Geophysical Laboratory. He first took his investigations into minerals and the origin of life to NASA's Astrobiology Institute in 1996, where he continues to play an important role in the Astrobiology program.

His friend, paleontologist Niles Eldredge, attests to Bob Hazen's talent as a trumpeter (he was president of the MIT Symphony Orchestra in the late 60s). Hazen has appeared as a soloist with the National Gallery Orchestra, the Boston Symphony Esplanade Orchestra, at the Kennedy Center and on *BBC-TV* and performed with the Metropolitan Opera, Boston and National Symphonies, Orchestre de Paris and the Kirov and Royal Ballets (partial list).

Hazen is not infrequently on television and radio, most recently in the History Channel's *Origins of Life* documentary earlier this summer. He served on the Committee to revise the

National Academy of Sciences' publication *Science, Evolution and Creationism* (the book has been criticized by some for promoting a dying theory of evolution).

He lists two pages of awards and honors on his C.V., including National Science Foundation's Distinguished Public Lecturer (2007). Hazen has also served as president and vice president of the Mineralogical Society of America. His. B.S. and S.M. degrees in Earth Science are from MIT (president of MIT's Geology Club too) and his Ph.D. in Mineralogy and Crystallography is from Harvard.

Our recent phone conversation about astrobiology follows.

**Robert Hazen**: Of course astrobiology exists. Any human endeavor "exists" where there's a group of people, in this case it's probably 1,000 researchers, who have a common set of goals and aspirations...

Astrobiology is a search for the origin, distribution and future of life in the Universe. What its underlying assumptions and hypotheses are is quite clear. First is we know life exists on Earth. In some way there was an origin of life on Earth. As scientists we accept the hypothesis there's a chemical and physical basis for that origin that is natural, in accordance with natural laws.

**Suzan Mazur**: Astrobiology includes astronomy, biology, chemistry and geology?

**Robert Hazen**: Absolutely. And physics too. It's a very integrated science.

**Suzan Mazur**: That's a huge field.

**Robert Hazen**: It is a huge field, but it is also one fairly well defined in the sense that we have some very specific questions. We want to understand the origin of life.

**Suzan Mazur**: "What are its laws?" – as philosopher Jerry Fodor has asked.

**Robert Hazen**: Well, there's the epistemological. Does every scientific question have to be first rooted in laws? Here's my take on it.

I did a book some years ago with a distinguished biologist, Maxine Singer. It was basically exploring the unanswered questions in science. Rather than ask what are the laws, what we zeroed in on is the fact that there are four distinct kinds of questions scientists ask about the natural world.

*First* are existence questions – going out and reporting what's out there. The astrobiology field is looking for signs of ancient life on Earth as well as elsewhere.

*Second* is origins. Astrobiology asks, where did life originate?

*Third* is process. How do things work? Scientists spend most of their time thinking about how systems work. Astrobiology addresses exactly that: How do living things organize themselves? How did they evolve? How did they adapt to different environments?

*Finally* there are applied questions. Ways in which you can take your understanding of the first three categories and alter or improve human existence. That's been shown over and over again with astrobiology. For example, in the field of molecular evolution. Used extensively in medicine.

Astrobiology as a field has a core set of fundamental questions; that's mainstream to what scientists do. So the question: What are the astrobiology laws? is a little bit of a red herring.

The thinking is not that there are laws of astrobiology, so therefore it is a science, rather that there are fundamental questions astrobiologists are asking and attempting to answer through scientific processes – and therefore astrobiology is a science.

**Suzan Mazur**: Niles Eldredge earlier this year told me that you're a terrific trumpet player and mineralogist, but that you're not an evolutionary biologist. So "be careful" he said.

**Robert Hazen**: Niles is absolutely correct, I'm not an evolutionary biologist.

**Suzan Mazur**: But I'd like to ask you how scientists can comfortably start in the middle of the evolutionary process – that is, once life is already present, and make accurate assessments, if the connection to origin of life remains elusive? And that's where you come in as a mineralogist. Right?

**Robert Hazen**: Yes. Absolutely. And that's true.

**Suzan Mazur**: You're comfortable speaking to origin of life issues.

**Robert Hazen**: I'm comfortable speaking to origin of life issues because the origin of life is not a biological or evolutionary process. Origin of life is a geochemical process. It involves self-organization – where you've brought together rocks and minerals, the atmosphere and the oceans, various chemical reactions that occurred. Some of those processes are deterministic.

We can study them with great rigor in the laboratory. We see there are processes of self-organization – primarily on surfaces. And those surface organization processes you can study quite distinctly and they have been studied a great deal and continue to be a very exciting area of forefront research.

**Suzan Mazur**: But evolution doesn't stop after the mineral phase, does it?

**Robert Hazen**: When we talk about evolution, there are many many definitions.

**Suzan Mazur**: Well that's a problem. The semantics.

**Robert Hazen**: The semantics. I'm not talking about evolution by a biological natural selection genetic mechanism. What I'm interested in is the period before there was a genetic mechanism.

**Suzan Mazur**: But can you have a separation?

**Robert Hazen**: Oh you absolutely have to have a separation between the two, I think. The first step in the origin of life – the earliest chemical steps, which is what I'm studying.

Let me make it clear, when I say I'm studying the origin of life, my personal interest and my research is in the earliest stages of what's known as "chemical evolution" – the chemical synthesis of organic materials and the organization of those. This is not yet a life form in any sense of the modern word.

What we're looking for is the earliest stages of organization of those chemicals. So this is really a geochemical process of organization. And the reason that astrobiology tackles this is that we say that these early steps are going to occur on any Earth-like planet and moon.

**Suzan Mazur**: Complexity pioneer Stuart Kauffman said that natural selection exists throughout the Universe wherever there is life. Harvard's Andrew Knoll, who appeared in the *Origins of Life* History Channel documentary with you this summer, told me at the Rockefeller University Evolution symposium in May that "It's natural selection every step of the way."

However, Stuart Newman in his recent paper for *Physical Biology* proposes all of today's 35 or so modern animal phyla emerged as a result of self-organization by the time of the Cambrian explosion a half billion years ago using a pattern language (dynamical patterning modules), with selection following as a "stabilizer".

What is your position on natural selection?

**Robert Hazen**: Again, I think there's a semantics question here. Selection – I'm not talking about Darwinian natural selection in the way that Darwin characterized it as survival of the fittest of the population. That's very specific.

Just as evolution has many different meanings from change over time to common descent to complexification to the specifics of Darwinian biological natural selection, I think the word natural selection has that ring of Darwinian survival of the fittest, but there's also a more general use of the term selection. That is, that if you supply selective pressures, which always happens in a natural setting if there are gradients of energy. Or if there are cycles such as day - night, light - dark, hot - cold, wet - dry. And those cycles tend to winnow out certain chemicals and enhance the population of other chemicals. That is a selective process.

It is a natural process. It is not "natural selection" the way Darwin used it. But it is a natural selective process.

I agree with what everybody just said (above). I'm not sure that it didn't occur before 500 million years ago, because I think we definitely see natural selection going on in microbial populations. We see that today. My assumption – although I'm not an evolutionary biologist – my assumption would be that right from the very first cell that had a genetic apparatus and was replicating and was competing to survive in a whole variety of different environments, that you would have had some kind of logical Darwinian natural selection going on.

**Suzan Mazur**: So you don't think that keeping natural selection in the equation may prevent us from seeking life as it may exist elsewhere?

**Robert Hazen**: When we look for life elsewhere, it's really a chemical problem. The characteristics of all living things, no matter what you think they are – chemical systems that are imagined, ones that we haven't yet imagined – the common characteristics of all those different systems is going to be chemical idiosyncrasies.

What I mean by that is that when you have all the prebiotic organic molecules that could be synthesized – hundreds of different amino acids, left-handed amino acids, right-handed amino acids, lots of different sugars, lots of different lipids, etc. – life tends to use a very small subset of all those different kinds of molecules. So in searching for life elsewhere, it's going to be a search for distinctive suites of organic molecules. It's not a search for a structure or organization.

I think it's very unlikely, for example, that we'll see on Mars some fossil of a thing, even with a microscope. But we may find suites of preserved organic molecules. Like in petroleum where you find a very idiosyncratic suite of molecules. It's not the thing you expect if you blasted a synthesis out of lots of hydrocarbons.

**Suzan Mazur**: Are you looking at abiotic oil at all?

**Robert Hazen**: That is a very different subject, and there are many resources. There is a new book on oil by Eric Roston. I just chaired a conference at the Carnegie Institution, called the deep carbon cycle. It's on the Carnegie web site. We had experts from all over the world. You can see most of the lectures, including lectures by Russian scientists who believe that petroleum is virtually all abiotic. And hear lectures by American petroleum geologists who think oil is virtually all biological. It's still an unresolved issue.

**Suzan Mazur**: Are you familiar with the *Astrobiology Primer* that NASA/NAI put out that Lucas Mix edited, which only refers to natural selection and neutral selection? Isn't that a bit limiting?

**Robert Hazen**: So much of this is semantics. When you talk about selection, some people say no it's not selection, it's self-organization. But self-organization is a selection process. It's not natural selection in the Darwinian sense. The fact that one molecule is responding to its local environment is always a driving mechanism. The whole idea of self-organization of lipid vesicles, for example, of self-organization on a mineral surface – that is a selection process. And very much in the mainstream of what people are thinking about.

Don't be concerned about a conflict between people who say – "oh, it's self-organization" or "oh, it's selection". To some extent this is really a kind of a...

**Suzan Mazur**: Word game.

**Robert Hazen**: It's a word game. The words evolution and natural selection – they have this emotional baggage, which is understandable I think. But it doesn't have to be a big conflict.

Selection as a universal process starts with the Big Bang. Which isotopes form? What kinds of planets form? How does the core separate from the surface? Those are all selection processes.

**Suzan Mazur**: I saw your early work with Larry Finger cited in *Evolution without Selection* (1988), the book by University of Lund cytogeneticist Antonio Lima-de-Faria. It was in a discussion regarding origin of form in minerals, the atomic composition and assembly.

**Robert Hazen**: I wasn't doing origin of life work back then.

**Suzan Mazur**: Lima-de-Faria cites your work in relation to atomic composition and assembly in carbon and the crystallizing as diamond and graphite. He thinks minerals have had their own separate evolution, and that the mineral evolution preceded the biological.

He was saying this 20 years ago, but he's told me recently that his thinking has not changed. He says it's important to go back to the atomic, chemical and mineral footprints to get the story right about biological evolution – which he considers the "terminal" phase.

I actually brought this up at a World Science Festival panel on the "Laws of Life" at NYU in June. Synthetic biology pioneer Steve Benner and astrobiologist Paul Davies were on the panel. Rockefeller University president Paul Nurse hosted the previous panel and said he predicted that biology will be looking to physics for answers on evolution.

Benner said he agreed with his "distinguished colleague from Lund" Lima-de-Faria: "But certainly our view of how life originated on Earth is very much dependent on minerals being involved in the process to control the chemistry."

And then Paul Davies said, "There has to be a pathway from chemistry to biology, powerful levels before Darwinian evolution even kicks in."

Lima-de-Faria thinks "[N]othing essentially new arose as biological evolution emerged." He says what looks new are the "combinations that seem drastically unrelated, only because they are so severely canalized into a narrow and limited number of variation channels."

I was just wondering what your response to that is?

Are you still there?

**Robert Hazen**: Absolutely. I'm listening to you very intently.

**Suzan Mazur**: Lima-de-Faria says that minerals and simple chemicals like water don't have genes but "display the constancy of pattern and the ability to change by forming a large number of forms" – they behave like an organism. That neither water nor calcite have genes but "possess mechanisms… we consider fundamental gene attributes." He also writes that "the main types of plant and animal patterns are already present in minerals".

There was a rustle in the room when I brought up the ideas of Lima-de-Faria, but if the Benner and Davies comments are representative, more scientists are thinking this way.

**Robert Hazen**: I think I understand what he's driving at and I don't want to endorse or reject the points of view cause I'd really need to hear more in detail.

**Suzan Mazur**: You've never read the book?

**Robert Hazen**: I just finished a paper called "Mineral Evolution" that is now in press in a journal called the *American Mineralogist*. There are seven co-authors and we talk about this idea of the mineral kingdom going through an evolutionary process. By that I want to make very clear what I mean by evolution. It's not just change over time. In this case it's diversification or complexification over time.

Let me give you a brief abstract of this idea: All terrestrial planets like Earth, Mars and Mercury and their moons, etc., begin in a pre-solar cloud of dust and gas. In that pre-solar cloud, there are microminerals – about a dozen different minerals: diamond, graphite, corundum, spinel – all together about a dozen different minerals.

And as you clump those together to form the earliest bodies and meteorites and asteroidal bodies, you get about 60 different minerals through heating of the sun in the primary formation of minerals. About 60 different minerals that form the most primitive minerals in what are called chondrite meteorites.

And then you go through periods of aqueous alteration. And then these planetismals get larger. You get up to about 250 different minerals. You see increasing complexification, from 12 to 60 to 250 to 350 in a place like Mercury, to 500 in a place like the moon.

And with plate tectonics you add more minerals and you go to 1,000. You go to 1,500 and finally when life kicks in on Earth you get maybe another 3,000 known minerals. So about two-thirds of all the known minerals on Earth actually result in a very complex feedback mechanism with life. That's what we call "mineral evoluton".

Now it's not Darwinian natural selection by any means. But it is a change over time. And it follows fundamental laws of physics and chemistry – and selection, leading to a gradual complexification or diversification of the mineral kingdom.

That's something I see as a process of evolution, which in many ways, is parallel to biological evolution.

Niles Eldredge, who you mentioned, and I have been working on a paper called "Themes and Variations in Complex Evolving Systems". The idea being that in both natural systems and in non-biological and biological systems, in cornets – which Niles Eldredge talks about – in language, you see similar themes.

You have species. You have diversification. You have extinction. You have punctuation. You have selection. Those five characteristics and others as well are common to all evolving systems whether it be minerals or language or biology or microbes or bears. And the fact is that there are also fundamental differences amongst those different systems.

I'm very sympathetic to people who see echoes of biology in mineralogy or echoes of biology in language.

**Suzan Mazur**: Here's something else that Lima-de-Faria said about minerals. "The body of a human like that of any other mammal is built according to a crystal plan. Bilateral symmetry to humans is indistinguishable from the twinning process in minerals."

**Robert Hazen**: Hmmm.

**Suzan Mazur**: "The two halves of the body are built just as contact crystal twins. They are intercharacterized by the fusion of two structures that are identical. They have an axis in common and one half is grown in position which corresponds to a rotation of 180 degrees thereby becoming a mirror image of the other."

**Robert Hazen**: I understand the rhetoric. I think there are some inaccuracies in that particular statement. If you look at the internal organs, they're not all bilaterally symmetric.

**Suzan Mazur**: He's saying that the cell was formed by the same atoms in minerals and that the cell continues to receive many of them from this source. He notes that "it's not surprising that periodicity is present at the biological level." Things are ordered. The biology behaves according to the previous footprints that were laid down.

**Robert Hazen**: I certainly agree that the chemical principles that govern rocks and minerals are the exact same chemical principles that govern all living things. All molecules. All living cells. And the interaction of those molecules and biological systems. And that leads to chemical bonding from just a few basic types. And it leads to self-organization of those molecules into larger structures based on energetics. That's absolutely true. It's what governs the forms and minerals in the natural world. Because an individual molecule can only respond to its immediate chemical environment. An individual cell can only respond to its immediate chemical environment. So how you go from a single cell in development through the whole amazing developmental process that leads to a complex individual can only be governed by the local immediate chemical forces and mechanical forces that surround an individual thing and that leads to kinds of self-organization. And so what I would say is that there's a little more nuance.

The more nuanced view might be that, yes indeed, the chemical laws are the same for minerals and for living things. And in each case the local molecules and atoms respond only to chemical environment. However, I would say the biological system is different from minerals. And there are two fundamental differences between a biological cell or organism and mineral.

In a biological cell or organism there are genes that subtly change through mutation. As a result the chemicals involved can change. Quartz is always quartz. It's always SIO2. The quartz that formed on Earth four and a half billion years ago is the same quartz today. Whereas, a single cell with its genetic complement, which undergoes these mutational changes – presumably the mutations themselves are random, while the selection process is not – so gradually a cell can change from generation to generation. As a result the chemicals that the cell produces change. Now the same chemical principles operate on the cell as they do on the minerals.

**Suzan Mazur**: Lima-de-Faria (a cytogeneticist) would argue that everything is ordered. All these things respond to a certain order that's been established through the various evolutions (atomic, chemical, mineral, biological).

**Robert Hazen**: Again, I'm not an evolutionary biologist but I do teach introductory genetics in my science literacy class and I think it's very much against the mainstream point of view.

We can sequence your cytochrome, mine and those of a chimpanzee and you can see that there are small characteristic differences and either those were created by design differently, which I don't accept cause it's not a scientific explanation, or else there was a mutational process that gradually caused different letters of the genetic code to change. Therefore, different amino acids to be introduced to the protein and therefore those proteins are different chemically, which means that they have different chemical characteristics in terms of their response to their immediate neighbor. That's the only thing that can really control the behavior of a biological system or a mineralogical system is the local chemical forces that are created by specific atoms and molecules.

If you accept a natural explanation and how the natural world works, then you basically have to talk about interactions among atoms and molecules. And self-organization is a consequence of that. But if you say it's a chemical process, then you have to be concerned about what those chemicals are. Inasmuch as chemicals can change in a biological system through gradual mutation and therefore genetic changes, different proteins form because there's a different sequence of amino acids. It's different with minerals.

**Suzan Mazur**: In light of what Paul Nurse said at the World Science Festival panel that biology may now be looking to physics for answers regarding evolution and your review of Stuart Pivar's concept of the toroidal model. By the way, he's just come back from Santa Fe Institute where he apparently had an enthusiastic reception about his work.

**Robert Hazen**: Is that someone from the Santa Fe Institute or Stuart who told you that?

**Suzan Mazur**: He met with a panel of people there, Geoffrey West and others.

**Robert Hazen**: Have you spoken with them?

**Suzan Mazur**: That's all I've heard.

**Robert Hazen**: When I reviewed Stuart Pivar's book I tried to take it as seriously as I could. I gave him a four and a half page detailed critique. http://www.scoop.co.nz/stories/HL0807/S00339.htm. Said this was never to be used except in its entirety. He extracted four or so paragraphs from that – put in juxtaposition – made it sound like I was endorsing his work. Or at least taking it seriously. And I really want to make it clear that I am not in any way a supporter of Stuart Pivar's work. I think there are aspects of his ideas that could be tested. That it's very interesting the idea of where different morphotypes come from. And commonality – that's a fascinating idea that could be explored. But his rhetoric rejecting out of hand Darwinian evolution. . . .

You see, let me take a step back. There's one thing I want you to understand about where I come from. It's more philosophical. It's that the history of science – as you have found out and you're very good at picking up on this – is often laced with people who establish a position that is very intransigent. This is the way it is. And someone else will say no this is some other way.

You have people who, for example, say it's gradualism versus punctuation. It is neptunism versus plutonism. It's uniformitarianism versus catastrophism. And these polarized debates have popped up over and over in the history of science.

It's a way that someone can make noise. It's a way that they can establish themselves or take a stand or somehow get behind a particular idea and it gives their career a focus. But what I'm finding over and over again in the history of science is that these dichotomies are false. That it's a much more nuanced answer.

It's not uniformitarianism or catastrophism. In fact, the history of Earth is characterized by long periods of gradual changes, punctuated by very dramatic change, like asteroid impact. Both are true. It's not neptunism versus plutonism. Rocks form both by agencies of water and agencies of heat. It's not just whether the early atmosphere was nonoxygenic or oxygenic. There was a gradual change from one to the other.

In the case of Stuart Pivar, he is promoting an idea about the structural organization of biology which I think has some very real merit, but he rejects out of hand anything to do with Darwinian evolution.

**Suzan Mazur**: A lot of people do.

**Robert Hazen**: People are polarized. It's not necessary to reject one in order to accept the other. The agency of change in biology over time that we see in the fossil record, that we see in modern life, that we see in microbes in hospitals, that we see in viruses much more rapidly have to do with mutation and accumulated changes. The changes themselves may be random, but the selection process is not. Selection is never random. By definition selection is non-random. In fact, in many cases it's almost a completely deterministic aspect of change.

**Suzan Mazur**: How fast is astrobiology growing?

**Robert Hazen**: Astrobiology is growing tremendously because there is a stable source of funding. Let's face it. Science is a social endeavor. If people can get jobs, they're going to go into the field. Right now NASA and other government agencies and also non-governmental agencies are putting money into this. They see this as a very exciting and promising field. We're also learning things about the natural world, about the extremes of life that have tremendous technological implications. And those technological implications help drive science as well.

**Suzan Mazur**: Do you have plans to update the content of the History Channel *Origins of Life* documentary which I think identified Darwinian natural selection as the mechanism of evolution?

**Robert Hazen**: It first appeared in June so I don't think they're going to do anything right away. You know how these things are. People move on to other projects... There are always new shows on origin of life coming along because there are a lot of cable channels and they all have to fill up 24 hours a day.

**Suzan Mazur**: We haven't seen these kind of topics emerge on *Charlie Rose* yet or serious television panels.

**Robert Hazen**: And the trouble with talking to me as a scientist – I like to bring in nuance. I don't like black and white. That's why as much as I have strong opinions about things like intelligent design, I don't do debates. Public debates. Because it's just the wrong forum.

**Suzan Mazur**: Can you address the difference between self-organization and intelligent design. There seems to be a reluctance to talk much about self-organization because of a misperceived connection to intelligent design. In fact, the word self-organization as part of the new Extended Evolutionary Synthesis, which the Altenberg 16 kicked off earlier this month in Austria, has been tucked into the umbrella term "phenotypic plasticity."

**Robert Hazen**: Maybe that's true [reluctance to talk about self-organization] in some cases, but not for me because I've seen self-organization at work in test tubes, seen it at work under a microscope. Self-organization is just a response to local chemical interactions, chemical bonding. There's nothing mystical about it. There's nothing intelligent or designed about it.

I think there's also a semantics problem here. All the scientists I know accept the basic laws of chemistry and physics. Some of those laws have to do with the interaction of molecules. They're electrostatic in nature and they form metallic bonding, ionic and covalent bonding, hydrogen bonding, Van der Waals forces and so forth. Those kinds of ways that atoms interact are fundamental to the cosmos. They're built to the very nature of the electron.

So when you say, do you have to start applying principles of physics as opposed to chemistry, I see it as continuity. I don't see a division between physics and chemistry. That chemical bonding is a physical process.

Self-organization is just a consequence of minimizing the energy of a system which happens spontaneously. It happens inevitably. It's a deterministic thing. So how could self-organization not be a part of biology?

**Suzan Mazur**: What about self-assembly?

**Robert Hazen**: Self-assembly is the same thing. Semantically self-organization and self-assembly are the same – how big is the module? Self-assembly: does that refer to a bigger piece than self-organization – I don't know. To me they're pretty much synonymous terms.

**Suzan Mazur**: Some people see a distinction.

**Robert Hazen**: Maybe they're just doing it on scale. You can talk about the self-assembly of magnets...

**Suzan Mazur**: Right and as Lima-de-Faria points out if you take a hydra – a living organism and push it through a sieve, it will reassemble.

**Robert Hazen**: The same thing is true of a Golgi apparatus. It's the same set of principles. It's still physical and chemical interactions at the local scale – whether it's a piece of magnet or piece of the hydra or piece of the Golgi apparatus or lipid molecule.

**Suzan Mazur**: Not enough of this literature has been around in the media. People don't understand, even the biologists don't understand what self-organization is. There's also been an attempt to block the literature like by Eugenie Scott's National Center for Science Education. She's told me they don't support self-organization literature because people confuse it with intelligent design.

**Robert Hazen**: I saw an article just last week in *Science* or *Nature* about self-organization.

**Suzan Mazur**: That might have been the subscriber-only *Science* magazine article rereporting what I've covered over the last several months about Altenberg, but getting it wrong. You don't see self-organization being talked about in the popular media...

**Robert Hazen**: Well, you're the kind of person who could really build bridges here. And I realize it makes good press to point out where scientists have fundamental disagreements, but so often – as you say – it's just a matter of communication. I don't know any scientist who argues against the importance of local, chemical, physical interaction. Whether it be at the level of the atom, the molecule, the cell. Those local interactions lead to what I refer to as self-assembly or self-organization.

I've seen it talked about because I'm very much into the "synthetics biology" community where there has to be self-organization. You don't basically take molecules and put them together piece by piece with glue. You have to put them in a test tube and let them make the cell for you.

And that's what Jack Szostak (at Harvard) sees and that's what Dave Deamer (at University of California, Santa Cruz) sees and you can go down the list of the people who are doing this very basic synthetic biology work. Self-organization works.

Now I'm not saying that's replicating the origin of life on Earth. But if we're going to make synthetic life in the laboratory in the next 10 years, it's going to be done through self-organization.

**Suzan Mazur**: Well thank you for sharing all of this with me.

**Robert Hazen**: The kinds of things you're reporting on, you're reaching a wider audience. It is important in writing to highlight where there may be controversies or disagreements because that is what's most exciting, that's what moves science.

**Suzan Mazur**: Evolution can't just be an American perspective, can it? What do we hear from the Japanese, for instance, about evolution? There was a group of structuralists for a while in the 1980s called the Osaka Group. Molecular biologist Atuhiro Sibatani, Kiyohiko Ikeda, Pegio-Yukio Gunji. Not all of the group were Japanese, however, I-SIS's Mae-Wan Ho participated, Brian Goodwin, Gerry Webster, Antonio Lima-de-Faria, Giuseppe Sermonti, Dave Lambert, Leendert Van Der Hammen, Vladimir Voeikov and others were a part of it.

**Robert Hazen**: I hate boundaries. I hate disciplinary boundaries. Talking about physics versus chemistry. I don't see a distinction. I hate international boundaries. I hate when people set up walls.

We've got to draw the circle wider. Even the idea of a professional scientist versus a knowledgeable reader. There's a continuum here. And we're all part of this search for trying to understand where we come from and who we are.

**Suzan Mazur**: Thanks so much for sounding the trumpet!

# Roger Buick & NASA: Follow the $H_2O$ or Energy Not Selection

*July 27, 2008*
*9:14 pm NZ*

Roger Buick, a native Australian, not only looks like a "rock star" with his long dark hair swept to one side and has the name to match, he actually is one. That is, he studies rocks and evidence of sulfur-eating bacteria and pre-Snowball Earth eukaryotes and such, while thinking about the possibility of life on Mars. Buick is the University of Washington professor of Earth and Space Sciences and Astrobiology, who in 2001, along with Yanan Shen and Donald Canfield, found the oldest visible evidence on Earth (in Australian rock) for a specific metabolic life process. When you see him – we met at the Rockefeller University Evolution symposium in May where his lecture was the crowd pleaser – you get the sense he's been somewhere you'd definitely like to go.

During the Rockefeller symposium cocktail hour, Buick informed me I'd upset the carefully delivered lecture of Harvard's Andrew Knoll ("it's natural selection every step of the way") by introducing a question from the floor about Stuart Newman's hypothesis that the 35 or so modern animal phyla self-organized by the time of the Cambrian explosion a half billion years ago without a genetic program.

In addressing the American Astronomical Society's annual meeting in Seattle in 2006, Roger Buick made this assessment of the budding field of astrobiology:

> "We don't know much yet, but it's going to be a huge amount of fun finding out. . . And everyone has something to contribute."

Buick has a Ph.D. (with distinction) in geology and geophysics, and a B.Sc. (Honors 1st class) in zoology and geology from the University of Western Australia. He also lectured at the University of Sydney's School of Geosciences (tenured). He was a postdoc fellow at Harvard. For several years along the way he worked with Sipa, BHP and other mining companies as an exploration geologist.

Our phone interview about astrobiology follows.

**Suzan Mazur**: NASA mineralogist Robert Hazen mentioned to me yesterday that there are about 1,000 researchers in the NASA astrobiology program. Are you involved in the program and in what capacity?

**Roger Buick**: Yes. I've got grant funding from the NASA Astrobiology Exobiology and Evolution Program and I'm also affiliated with the NASA Astrobiology Institute through the University of Arizona. And when the University of Washington was one of the NAI teams, I was involved in that too. They're currently not funded.

**Suzan Mazur**: They're not funded?

**Roger Buick**: The University of Washington team isn't affiliated with NASA Astrobiology Institute anymore. There's a team called the Virtual Planetary Laboratory, which is a very dispersed team, and the Principal Investigator on that team, Vikki Meadows, is at the University of Washington but nobody else at the University of Washington is supported through NAI.

**Suzan Mazur**: So your lab at the University of Washington is not supported.

**Roger Buick**: Not by the NASA Astrobiology Institute.

**Suzan Mazur**: Can you talk about your lab?

**Roger Buick**: Yes. We have a great lab. Four of us from different departments have collaborated to build the lab. We now have five mass spectrometers and we can analyze stable isotopes from rocks, liquids, gases for about five or six different light elements. So we can look at oceanographic processes, paleoclimatic processes, atmospheric processes. And also deep time. Biological evolution through biological isotopic fractionation within old rocks.

**Suzan Mazur**: I'd like to get into more of your work later in the interview, but getting back to the funding aspect of this story – Bob Hazen also said that a lot of money is being put into astrobiology not only from NASA but from other government agencies. I was wondering if you had any concern about academics being co-opted into the program because that's where the money is? I mean even the American Philosophical Society is partnering with NASA to provide grants.

**Roger Buick**: I don't think academics ever get co-opted into anything. But they do tend to follow the money. There's no coercion in it. Academics are greedy for cash like anybody else.

**Suzan Mazur**: But there's a genuine interest in the Astrobiology program, it's not just that that's where the money is.

**Roger Buick**: Oh yes. Before astrobiology was invented, people were interested in astrobiology. You just have to look at the student interest in astrobiology. They're not following money. They have very little clue about what research money does for science. On the undergraduate and graduate level, there are a large number of students who respond enormously to anything astrobiological.

I teach a lot of courses that aren't astrobiology, but whenever I throw something astrobiological into one of my nonastrobiological courses – it's the part of the course that really grabs the student. You can see them light up!

**Suzan Mazur**: When was astrobiology invented? And what does it consist of – what areas?

**Roger Buick**: It consists of almost everything. It was invented in about the late 90s.

**Suzan Mazur**: 1998 or something.

**Roger Buick**: Yes. Some time around there.

**Suzan Mazur**: By the way, I had a subsequent conversation with philosopher Jerry Fodor regarding his comment to me earlier this year that "Astrobiology doesn't exist. What are its laws?" He's updated his remarks and now says the following: "I did?" Fodor said he doesn't know the field.

**Roger Buick**: But the interest has been there since Jules Verne and H.G. Wells. That's what they were writing about.

**Suzan Mazur**: If Fodor says he doesn't know astrobiology, there just may be others...

Do you think that the private sector – scholars, for example – should be able to share in a piece of the NASA pie? I think that of the $17.5 billion 2008 budget, something like only $200 million was requested for innovative partnering programs for small businesses, etc. Proportionally a tiny amount.

**Roger Buick**: I'm not in favor of any government subsidy to private industry groups.

**Suzan Mazur**: What about private scholars?

**Roger Buick**: There are very few private scholars in the United States who have the infrastructure with which to be able to participate in astrobiological research in a big way.

**Suzan Mazur**: Isn't that un-American?

**Roger Buick**: Don't ask me what's American and un-American or you'll start me toward politics. Un-American is a very pejorative term and has had a disastrous history in United States politics. I don't think anyone should ever mention the phrase "un-American". I think of the House Committee and all that sort of stuff.

**Suzan Mazur**: You and your colleague Birger Rasmussen are credited with discovering the world's oldest oil, which you found trapped between mineral grains of rock 3.2 billion years old.

**Roger Buick**: That's correct.

**Suzan Mazur**: Have you approached Sotheby's?

**Roger Buick**: No. Even at current oil prices, I think the value of a nano-barrel of oil is infinitesimally small.

**Suzan Mazur**: Then in 2001, along with Yanan Shen and Donald Canfield you found what's considered the oldest life on Earth in Australian rock dating to about 3.47 billion years old?

**Roger Buick**: It's not exactly the oldest evidence of life. It's the oldest evidence of a specific metabolic process carried out by life. There are claims for older life in older rocks elsewhere in the world. But what we found was the oldest evidence for specific metabolic style, which implies the organisms responsible were as sophisticated in their biochemistry as modern organisms. It's a claim for modern-style life. . .

The work with Shen and Canfield was discovering evidence for sulfur-eating bacteria and immediately overlying that there are stromatolites. We don't know what sort of bugs made them. They could well have been bacteria using sulfur gases for photosynthesis.

**Suzan Mazur**: There was an NAI report in 2001 that said:

> "Buick says that the presence of sulfate-reducing bacteria almost 3.5 billion years old suggests that a wide range of microrganisms has already "colonized the early Earth" forming a rudimentary food chain."

You're quoted in that same article saying:

> "From spectral analysis we know that there are lots of sulfate minerals on the surface of Mars. If that planet was warmer, wetter and inhabited more than 3.5 billion years ago, we might be able to find older signs of biological sulfate reduction there provided of course that NASA sends a bloody good field geologist with lots of experience of particularly ancient rocks in remote places."

It sounds as if you have no doubt that life originated elsewhere in the Universe and colonized Earth or were you misquoted?

**Roger Buick**: No. Not misquoted. But what I was trying to get across is that if life had started on Mars, we might have a better chance of tracing its earliest evolution on Mars than we would on the Earth because there are very few rocks on the Earth from the first billion years of Earth history. And the 3.47 billion year old rocks from Australia that we were working on are pretty much the oldest ones that you can find on the Earth that would show evidence of this sort of metabolism. But on Mars, there's a good geological record from the first billion years of its history. So Mars might be a better place to explore for how life started and how it initially colonizes the planet than the Earth, if there was ever life on Mars.

**Suzan Mazur**: Also, I sense that your appearance at the Rockefeller Evolution symposium in May discussing this subject is an indication that the scientific community seems to be siding with you and not UCLA's Bruce Runnegar. Runnegar didn't speak at the Evolution symposium. His challenge is that the sulfate in the Australian rock was reduced not by bacteria but from exposure to hydrothermal fluids. Could you comment about that? Is that challenge still there?

**Roger Buick**: As far as I know the data has never been published except in conference abstracts.

**Suzan Mazur**: Whose data?

**Roger Buick**: Runnegar's. I think the debate has moved on substantially since he was making his comments. There's a lot of new data.

**Suzan Mazur**: Supporting your argument and your find?

**Roger Buick**: Consistent with our find and inconsistent with a hydrothermal origin for that fractionation. You can look at other isotopes of the element sulfur and show that the fractionated sulfur that we found has been through atmospheric processes which would argue against a hydrothermal cause for any isotopic fractionation.

There's an important paper in *Nature* by a French group suggesting that not only was microbial sulfate reduction active but also microbes were disproportionating elemental sulfur from those rocks. It that's correct, that would bear out my contention that there was quite a diverse microbiota living in that environment very early.

**Suzan Mazur**: Have you had significant dialogue with NASA regarding your involvement in a Mars investigation?

**Roger Buick**: It's been suggested to me that I might be a good person to assist in a site selection for the next Mars Rover mission. Apart from that, no. I haven't been directly involved in Mars research.

**Suzan Mazur**: On the subject of crystals. This is an area you can speak to?

**Roger Buick**: It depends on what aspect.

**Suzan Mazur**: On the subject of mineral evolution in relation to biological evolution, Antonio Lima-de-Faria wrote in his classic 1988 *Evolution without Selection* – he's a cytogeneticist from the University of Lund. He thinks we're in the fourth level of evolution, the biological, which was preceded by the atomic, chemical and mineral and that evolution from minerals to living organisms used four different routes: solid crystalline, liquid crystalline, quasi crystalline and the amorphous.

He points to Shechtmanite, an alloy of manganese and aluminum, displaying a five-fold symmetry that was previously considered to occur only in living organisms. He also thinks life has no beginning, that it's a process inherent to the Universe. Says we've never had a theory of evolution. Can you comment on this?

**Roger Buick**: There are a number of workers who consider that interactions with minerals and mineral crystals was a significant component of prebiotic chemistry. And those ideas have been around for 50 years I would guess. The first person I can think of is Graham Cairns Smith back in the late 60s, early 70s, proposing that clay crystal surfaces were important for prebiotic chemistry.

A wide range of organic chemical/mineral crystal surface interactions have been proposed as significant in the origin of life. Personally, as a geologist I'm quite drawn by those models. The experimental work that has been done indeed shows that mineral crystal surfaces can assist in plausible prebiotic chemical reaction.

But I'd be surprised if it were the case that crystal surface chemistry is a complete explanation for the origin of life. The world's a complex place and to try and pin the origin of life on one particular sort of process is a bit presumptuous.

**Suzan Mazur**: In our search for life elsewhere in the Universe are we looking for the right thing if we continue to insist it happened as a result of Darwinian natural selection?

**Roger Buick**: I don't think that the astrobiological search for life elsewhere necessarily presupposes a Darwinian natural selection model for the origin of life. The NASA mantra is: Follow the water or follow the energy. It's not follow the selection.

I think most astrobiologists are reasonably agnostic about how the origin of life occurred.

**Suzan Mazur**: It's interesting that you're saying this because the NASA/NAI -supported *Astrobiology Primer* – I don't know if you saw that.

**Roger Buick**: Yes.

**Suzan Mazur**: The editor-in-chief is an Episcopalian priest. Natural selection was the mechanism of evolution cited and there was a small section on neutral selection. I spoke to the editor about this and he did say that there would hopefully be an update of the *Primer* in the next couple of years.

**Roger Buick**: I don't think that particular volume guides NASA space missions for this or that planet or for signs of organized life or NASA's research agenda in astrobiology. **That volume is meant to be for basic education** [emphasis added].

**Suzan Mazur**: Meant for basic education? I think that's a problem. The *Primer* editor has a major book on astrobiology coming out next year published by Harvard University Press.

**Roger Buick**: Hmmm.

**Suzan Mazur**: I was wondering also, since you discovered the oldest oil, if you could comment on the possible existence of abiotic oil that's now being talked about a lot.

**Roger Buick**: It's an argument that's been around for a long time. Thomas Gold wrote about it.

**Suzan Mazur**: I know Bob Hazen hosted a conference on it earlier this year at the Carnegie Institution. He mentioned that the Russians were very much behind the existence of abiotic oil. Is the idea of abiotic oil simply a way of prolonging the life of the oil industry? Or is abiotic oil a reality?

**Roger Buick**: We know that there are several geochemical processes that can synthesize complex hydrocarbons out of very simple molecules and we also know that some of those processes may have been more active on an early Earth that had greater rates of volcanic activity, hydrothermal alteration and things like that.

So it's at least plausible that the 3.2 billion year old oil we found did in fact have an abiotic origin. We can't prove it one way or the other.

**Suzan Mazur**: Why not?

**Roger Buick**: We can't chemically analyze the oil because it's in such minute quantities. But we can go to slightly younger rocks and we can tease apart that oil in those and find out what molecules it's composed of. And when we go to rocks 2.4 billion years ago (something we published last year and the beginning of this year), you can analyze that oil in detail and you find molecules that could only have been produced by living organisms. Really complex multi-ring hydrocarbon molecules.

And if you go to every oil deposit we know of on the Earth today and you analyze that oil – it also has these biological indicator molecules. Abiotic oil might be produced now. It might have been more significant in the past. But it's not a significant component of any oil reservoir that we know of on Earth. And geochemistry shows that quite conclusively.

**Suzan Mazur**: So it's not any kind of quick fix.

**Roger Buick**: It's no kind of quick fix. But the fact that we can synthesize complex hydrocarbons out of simple molecules using geochemical processes suggests that you could in fact officially manufacture oil. But it would be at a very high cost. Whether it would be an economically viable substitute for biologically derived oil pumped out of the ground is another question.

**Suzan Mazur**: A final comment?

**Roger Buick**: Astrobiology's great fun. It's stretched me enormously. And I've loved it.

# David Deamer: Line Arbitrary Twixt Life & Non-Life

*September 10, 2008*
*4:42 pm NZ*

When I reached origins of life investigator David Deamer by phone at his lab at the University of California, Santa Cruz, he told me the NASA Astrobiology Program he's part of encourages public outreach, since the program is publicly funded, and that he'd be happy to do an interview. But in the next breath Deamer revealed that the NASA Astrobiology program had no funds: "There's no money available to send out any new grants at all."

It was David Deamer's spelunking adventures growing up in Ohio that first sparked his curiosity about origins of life. By 1957, he was recognized in a Westinghouse Science Talent Search for his investigation of the self-organization of protozoa. He says "Ilya Prigogine's pioneering of complexity was an inspiration – for us all."

A half century and many awards later, David Deamer is Professor of Biomolecular Engineering and Research Professor of Chemistry and Biochemistry at the University of California, Santa Cruz where he directs a lab on self-assembly processes and the origin of cellular life. The lab has been supported for over 20 years by the NASA Exobiology program and for over 10 years by the National Institute of Health.

He is also part of the Carnegie Astrobiology team affiliated with the Carnegie Institution in Washington, which has been investigating "Astrobiological Pathways: From Interstellar medium, through Planetary Systems, to the Emergence and Detection of Life". And he is informally associated with NASA Ames Research Center.

Deamer was a professor of zoology at University of California, Davis for more than 25 years before coming to UCSC. He has chaired many academic departments at UCSC and UC, Davis (Zoology, Biophysics, Biomolecular Engineering) as well as conferences – including a NATO Advanced Research Workshop in Hungary: "Polymers in Confined Spaces".

In 1988, along with musician Susan Alexjander, Deamer put DNA to music to make "microtones".

He has six patents and is the author of 126 peer-reviewed papers as well as 10 books, including *Being Human: Principles of Human Physiology, The World of the Cell, Origins of Life: The Central Concepts* – with another forthcoming (2010) from the University of California Press: *Stars, Planets, Life*.

Last year David Deamer lectured at the "What is Life" Symposium in Kyoto, Japan and the year before that on "Self-assembly processes in the prebiotic environment" at the Royal Society in London. He will be a featured speaker again on self-assembly at the upcoming AAAS meeting in Chicago.

Deamer's Ph.D. is in physiological chemistry from Ohio State University School of Medicine and his B.S. in chemistry is from Duke. His current research involves how DNA can make its way through nanoscopic pores in membranes.

Honors include Fellow, International Society for the Study of the Origin of Life (2002), Distinguished Lecture series, Graduate Center, CCNY (2004), Distinguished Lecturer, Royal Society of New Zealand (1989), Guggenheim Fellow (1986) and the Westinghouse Science Talent Search (1957), among others.

He lists a dozen public service roles on his C.V. He currently serves on the editorial boards of the *Journal of Bioenergetics and Biomembranes, Astrobiology* and *Origins of Life and Evolution of the Biosphere*. He has previously served on the NASA Space Science Advisory Committee, NASA Astrobiology Roadmap

Panel, and chaired the NASA Panel on Exobiology (1991-1995).

My phone conversation with David Deamer follows.

**Suzan Mazur**: The scientific establishment and the mainstream media are slow to accept that there are mechanisms involved in evolution beyond Darwinian natural selection. Part of the problem is that they are unclear what these other mechanisms are. Can you tell me, for example, what the process of self-assembly is and self-organization and how they differ from one another?

**David Deamer**: It would be good to have more precise definitions because I've tended to use the terms more or less as synonyms. Let's start with self-assembly, which I define as a molecular process that produces ordered structures from disordered components, yet is energetically downhill, in the sense that an external energy input is not required to get it to happen.

In contrast, most life processes are energetically uphill. A source of external energy is required for polymerization of amino acids into proteins, which is the main growth process of life. Self-assembly is more like what happens to soap molecules in solution. Do you want me to go into technical detail?

**Suzan Mazur**: If you could describe these terms so a general audience can understand the science without being too technical, that would be great. The concepts now are ignored and dismissed as magic, "woo-woo," because of the spontaneous way they happen.

**David Deamer**: Okay, let's talk a little more about self-assembly. It is extraordinary what certain kinds of molecules can do in an aqueous environment. The example that I use is soap molecules in water. A soap molecule is just an oily hydrocarbon chain with a hydrophilic, or "water-loving" group at one end. When they are in a dilute solution soap molecules float around at random and pay no attention to each other.

But if the concentration is increased, the soap molecules begin to aggregate into little clumps called micelles which are composed of a few hundred soap molecules each.

**What drives this is a law of physics that controls the way that water molecules interact with the hydrocarbon chains of the soap molecules** [emphasis added].

Everyone has heard that "oil and water don't mix." At a certain concentration of soap there is no more room for the hydrocarbon chains to fit into the water structure, so they begin to stick together in micelles with all the oily chains pointing into the micelle, away from the water.

Now let's add more soap. When we get up to a concentration we call the CVC, or critical vesicle concentration, the micelles begin to aggregate into membranes and the membranes form beautiful little vesicles.

These are microscopic versions of the soap bubbles that everybody has seen at the macroscopic level. But if you look at soapy water under the microscope, what you see are microscopic vesicles that form compartments with an interior volume that is separated by a membrane from the external environment.

The point is that the membranes of cells are also produced by self-assembly. Nothing in the genes tells a membrane how to be a membrane. Instead the genetic information in the genes tells the cell how to make the fatty acids (the scientific word for soap) and how to assemble the fatty acids into more complex lipids. The lipids then assemble spontaneously into membranes, the boundary structures of all living cells.

**Suzan Mazur**: Would you describe self-organization? Self-organization is an open system?

**David Deamer**: Yes. If self-assembly is a spontaneous, energetically downhill process, I would define self-organization as a step up from self-assembly in which more complex structures, including living organisms, use energy to organize themselves into functional aggregates.

**Suzan Mazur**: You say it's a step up. So you see it as some sort of...

**David Deamer**: Increase in complexity.

**Suzan Mazur**: Are you saying there's a connection between self-assembly and self-organization?

**David Deamer**: It's analogous to the connection we might make between inorganic chemistry and organic chemistry. Organic chemicals can be much more complex than simple inorganic chemicals. Self-organized systems are more complex than self-assembled systems and can even include populations of organisms that organize themselves in the ecosystem.

**Suzan Mazur**: Cell biologist Stuart Newman told me in a recent interview that self-organization requires a "flux of matter or energy to keep the structure in place".

**David Deamer**: I would agree with that. In contrast, self-assembly is spontaneous, and depends only on the interactions between molecules and with the environment.

**Suzan Mazur**: Change of subject. Why does NASA promote natural selection as the only mechanism of evolution in its literature – for example, in *Astrobiology Primer*, whose editor is a priest, and on television in the program *Origins of Life*?

**David Deamer**: NASA is speaking to the general public. They're just trying to keep it simple and explain evolution to people who may not know much about it.

**Suzan Mazur**: But there are other mechanisms contributing to evolution. The public is not being told about this. Not informing the public is not really serving the public.

**David Deamer**: The *Astrobiology Primer* and the *Origins of Life* program are intended for a **lay audience** [emphasis added]. Biologists agree that life started simple and became more complex through a natural process, and at the most general level we call that process evolution.

If I were teaching an advanced class in evolutionary biology to a college level audience, they would have enough preparation to deal with the other aspects that go into the evolutionary process beyond Darwin's initial explanation. It takes a lot of background to understand the details that contribute to the evolutionary process.

For instance, the Altenberg 16 you have written about are professional biologists who are trying to go beyond the simplistic explanations that involve nothing more than natural selection. They are bringing to the table ideas that require considerable knowledge to understand their argument.

I certainly wouldn't want to state that natural selection is the only process driving evolution, but if I am going to explain what that means, my audience needs to have enough information to understand the questions that are being raised.

**Suzan Mazur**: But as Stuart Newman, one of the Altenberg 16 scientists has pointed out, there would be more of an acceptance of evolution if the science was where it should be. He also says "old science" is being pushed in the mainstream media.

**David Deamer**: I get the point. Unfortunately, creationists have politicized the science so much that the very fact of evolution is being questioned.

Perhaps this is why scientists tend to fall back on the bedrock of Darwin's basic concepts when they speak in a public forum. No one denies the factual basis of evolution, but we are still learning how evolution takes place, particularly in animal and plant populations in ecosystems.

I have debated creationists and intelligent design people in public forums, and my impression is that they are not looking for scientific truth. Instead they are working to advance their political aim, which is to get Christian fundamentalism taught in public schools as an alternative to evolution.

**Suzan Mazur**: Cytogeneticist Antonio Lima-de-Faria from the University of Lund refers to the "cycle-of-submission" within academia where scientists are unnecessarily conservative, stick together, protect their foundation grants instead of recognizing the validity of alternative mechanisms and advancing the science. This kind of fundamentalism feeds a creationist perspective.

**David Deamer**: No matter what we do, the creationists are going to focus on things we don't know and forget about all the things we do know. I'm not sure there is any fundamental disagreement among scientists about the basic facts of evolution.

**Suzan Mazur**: There is clearly a horde mentality alive in the science blogosphere.

Should more scientific inquiry into the origins of life be encouraged by opening up the peer-review process? Often the papers of independent researchers are rejected because they're outsiders and may take an unorthodox approach.

**David Deamer**: I would like to see the evidence you cite that independent researchers are rejected because they're outsiders.

**Suzan Mazur**: They may take an unorthodox approach. In other words, they may not use all the science jargon that the scientific establishment is used to seeing in papers. So reviewers may reject a paper because it's not written with a tight science jargon. Rejection, you don't speak our language.

**David Deamer**: For every example you might give of a rejected unorthodox investigator, I could cite a counter-example. I'll mention just one, Gunther Wachtershauser, a Swiss patent attorney. Wachtershauser came up with an idea all of his own. He was an absolute outsider.

His idea was published first in 1988 in *Microbiology Reviews*. Because of the strength and the novelty of his idea and the elaboration that he was able to give to that initial publication, it really caught people's attention.

He followed this up with a *Scientific American* article and a series of other papers. I've been in meetings with Gunther. He's not one of the gang by any means, and yet we are paying attention and are testing his ideas. Some of them stand up to critical tests, others don't.

I would cite Wachtershauser as a clear example of an outsider breaking into the scientific process on the force of his ideas.

Then there are other independent researchers whose ideas just don't stand up to critical evaluation. They complain that they can't get their paper published, but the fact is that their ideas just don't make sense.

Peer review is the only process we have for sorting out the good ideas and getting them out there for others to think about.

**Suzan Mazur**: What is the standard for acceptance of a paper?

**David Deamer**: The standard is based on judgment calls by knowledgeable referees and editors.

**Suzan Mazur**: Acceptance doesn't require use of the same tight scientific jargon. It's essentially about a concept and clear thinking.

**David Deamer**: It's like being a good chess player. That's not a bad metaphor. A good chess player wins whether or not he or she is a member of a club.

If they come in and begin to win games based on unorthodox strategies, they are going to gain automatic respect. It's the same in science, which tends to attract people who think they have good and interesting ideas. I really doubt that there is a significant number of independent researchers who have really good ideas but are being rejected just because they are outsiders.

**Suzan Mazur**: Do you see any conflict of interest with many of the *Astrobiology* journal board members being NASA employees or NASA-affiliated?

**David Deamer**: It's a bit of a problem, but we deal with it. It comes down to numbers. There are 10,000 researchers who call themselves neurobiologists, and perhaps 30,000 - 40,000 chemists. But there are probably fewer than a hundred researchers who call themselves astrobiologists. Because their research is usually funded by NASA, it can be hard to find knowledgeable people to serve on editorial boards who don't have a perceived conflict with their NASA grants.

**Suzan Mazur**: But you're looking to include outsiders on the board?

**David Deamer**: Most of the board members are not NASA employees. There are eight senior editors, two of whom are civil servants at NASA Ames. There are 75 members of the editorial board, but only five NASA employees. Over a third of the board members are from other countries, so I think we are well represented internationally.

We certainly want to get more people into the Astrobiology program, and it is growing. There are now 500 people who attend the annual Astrobiology meeting, both younger researchers and people like me who've been associated with astrobiology since it began in 1996.

**Suzan Mazur**: How many academics are now in the Astrobiology program?

**David Deamer**: It's a few hundred if we include graduate students and post doctoral associates along with the principal investigators. Each member organization within the Astrobiology Institute has a principal investigator who assembles a small team of a few other faculty members that will get support from the program. And each faculty member might have one or two graduate students and a postdoc supported by the grant.

**Suzan Mazur**: And your affiliation with NASA at this point is…

**David Deamer**: I am associated with the Carnegie Astrobiology program that is funded through the Carnegie Institution of Washington. I'm also informally associated with the program at NASA Ames.

**Suzan Mazur**: You're not a NASA employee.

**David Deamer**: No. The only funds I receive through NASA are in the form of grants that typically support one postdoc and a grad student.

**Suzan Mazur**: So academics are not being lured into the Astrobiology program because of the money.

**David Deamer**: Definitely not! The research funds available are much less than grants from the National Institute of Health. They being lured into it because of the interest and excitement generated by this new field.

**Suzan Mazur**: The NASA Astrobiology funds are not expanding?

**David Deamer**: No. In fact, last year the budget was so restricted that new proposals could not be funded and were put on hold.

**Suzan Mazur**: Do we know very much about astrobiology 10 years or so on in the investigation?

**David Deamer**: Yes, absolutely. Astrobiology has put life on the Earth into a larger context of our solar system and our galaxy. The origin of life on Earth is likely to be a universal process, and that's why we are so excited by the discovery that Mars once had shallow seas. Perhaps in the next decade we will have clear evidence that life began there as well, by the same process of self-assembly that we discussed earlier.

It also has given us a vast amount of information about the history of life on the Earth. We now know that oceans were present well over four billion years ago, and there is evidence for life in the isotopic record that goes back about 3.8 billion years ago.

**Suzan Mazur**: Do you take a position as to whether life began outside Earth or on Earth?

**David Deamer**: I use plausibility arguments to answer questions like that. Plausibility is an individual judgment call based on knowledge. In my judgment it is implausible that life came to the Earth in the form of extraterrestrial spores or microrganisms. We can't rule it out but I consider it to be a very low probability.

On the other hand, I consider it to be very plausible that the organic compounds required for life to begin on the Earth were delivered to the Earth by comets and meteorites during late accretion, and that synthetic reactions were producing complex organic molecules in the early Earth environment. I think life most likely began on the Earth by a self-assembly process in which moderately complex chemicals self-assembled into vast numbers of microscopic encapsulated systems. **By a yet unknown process**, a very few of these happened to be able to capture energy and nutrients from the environment and began to grow by polymerization reactions. There is much more to the story, but this is my guess about how life began.

**Suzan Mazur**: What do you think the origin of the gene is?

**David Deamer**: I think genetic information more or less came out of nowhere by chance assemblages of short polymers. We don't know that these polymers were exactly like RNA and DNA of contemporary life, but in the laboratory we use those polymers as experimental model systems.

Most people are open to the possibility that there are simpler molecules that we haven't discovered yet that could contain what we now call genetic information. There may also have been specific sequences of monomers within a polymer that happened to allow it to fold into a catalytically active molecule. One idea is that RNA could have acted both as a catalyst and as a genetic molecule, so that at one stage in evolution life existed in an RNA world.

**Suzan Mazur**: So you see the line between life and non-life as being arbitrary?

**David Deamer**: Yes. There was probably an extensive mixing of genetic information at that time, as Carl Woese and others have suggested. This means that there was no tree of life at that time, instead just countless numbers of microscopic experiments occurring everywhere as the first catalysts and genes learned to work together in cellular compartments.

**Suzan Mazur**: Then does life have a beginning or is it just part of a process inherent to the Universe?

**David Deamer**: It's part of a process.

**Suzan Mazur**: Evolution starts when the Universe is born?

**David Deamer**: It depends on what you want to call evolution. The Universe is over 13 billion years old, but life originated on the Earth around 4 billion years ago. Biological evolution began with the transmission of genetic information between generations, and selective processes acting on variations within microbial populations.

**Suzan Mazur**: But can it be separated from the rest? Do you see biological life as relying on certain previous footprints?

**David Deamer**: Yes to both questions.

**Suzan Mazur**: Lima-de-Faria speaks of four levels of evolution – atomic, chemical, mineral and biological. He says there are coincidental patterns arising in organisms because they have the same atoms with the symmetries of the minerals transferred intact to the cell and organism level.

**David Deamer**: I agree to a certain extent, but there is still little evidence that minerals played an essential role in the process. Certainly astrobiology has given us a satisfying narrative of how life came to exist on the Earth, all the way from stellar nuclear synthesis to planet formation and habitability and then self-assembly of organics in aqueous environments. When energy sources impinge on these self-assembled structures they capture some of that energy and then interesting processes begin to emerge. So there's a narrative describing a continuum from which life gradually emerges.

**Suzan Mazur**: Do we have enough data to construct a periodic table in biology like that in chemistry?

**David Deamer**: I think we can construct a hierarchy of increasing complexity. It's possible to think of the periodic table as a hierarchy of complexity in which hydrogen is the least complex atom. As the elements become progressively heavier with the addition of protons, neutrons and electrons, each level has a different set of chemical properties and therefore a different set of potential complexities as they interact with each other. In this sense I think we could describe a hierarchy of complexity levels in life, but I don't think we would find much periodicity.

I'm writing a book about the origin of life for the University of California Press that is scheduled for publication in 2010. This is the approach that I'm taking in the book, that we can understand the origin of life in terms of the emergent properties of interacting systems of molecules.

**Suzan Mazur**: You've commented a little bit about the Altenberg group already – do you think that the Extended Synthesis is something the biology community will embrace at this point?

**David Deamer**: Epigenetic phenomena is one example of what can happen that is well beyond the usual idea that genes are all we need to understand evolution.

**Suzan Mazur**: Epigenetics has actually been tucked under the umbrella term plasticity in the Extended Synthesis. Do you think the Extended Synthesis was a good call, that the biology community will embrace this graft onto the modern synthesis?

**David Deamer**: This is how good science happens, when either an individual or a group of scientists think they might know something beyond the current consensus. They try to construct a new synthesis of ideas that has better explanatory power, and if they have a convincing argument, their peers will follow. Science should be open to these kinds of challenges.

This is what Steven J. Gould did with punctuated equilibrium, which caused so much controversy at first. Ed Wilson did this with sociobiology, and a consensus is slowly building that we can understand behavior in evolutionary terms.

**Suzan Mazur**: Along those lines, Stuart Newman predicts "a big turnaround in evolutionary theory". He cites non-linear and saltational mechanisms of embryonic development contributing to evolution. Newman has told me: "It was Darwin who said that if any organ is shown to have formed not by small increments but by jumps, his theory would therefore be wrong."

What are your thoughts about this?

**David Deamer**: I'd need to know more about this to have a knowledgeable comment. I don't know what Darwin really meant by his statement, or how it could be tested now that we understand so much more about embryological development.

# Ex NASA Astrobiology Institute Chief Bruce Runnegar

*October 2, 2008*
*4:06 pm NZ*

I have been having bits and pieces of communication with former NASA Astrobiology Institute chief Bruce Runnegar in recent weeks in between his field trips to Australia – sometimes via his wife Maria, a biochemist at the University of Southern California. Runnegar is investigating the oldest complex fossils – Ediacara fauna, in Australia as well as in Namibia, South Africa and Newfoundland. He speaks with a quiet ease and told me during my phone call to him last week at the University of California, Los Angeles, where he is a professor of paleontology, that he's "firmly disconnected from NASA for the last two years." But I was interested in how NAI's virtual organization works, so he agreed to explain. Runnegar is as comfortable discussing that as he is "the kerfluffle about the French Impressionists" on the Victorian scene, and defending natural selection in business (our chat was prior to the Dow falling 700 points).

Although no longer NAI chief, Bruce Runnegar still considers himself an astrobiologist. His interest is in events that coincided with the Cambrian explosion of multicellular organisms a half billion years ago. And while he recognizes self-organization as a mechanism of evolution, he doesn't buy the idea of plasticity in Cambrian multicellular organisms. Said Runnegar: "There was a common ancestor which didn't resemble anything that we see in the Cambrian."

Runnegar also challenges fellow native Australian astrobiologist Roger Buick's analysis that the 3.47 billion-year-old rock at Australia's North Pole holds the oldest evidence for a specific metabolic process carried out by life – sulfur-eating bacteria. Runnegar's got a year of sabbatical and plans to publish his results by December demonstrating that Buick's findings are not correct, that the sulfate deposits Buick found were due to exposure to hydrothermal fluids, not bacteria.

Bruce Runnegar served as the third director of the NASA Astrobiology Institute, beginning his tenure at Ames Research Center in September 2003. He was director of UCLA's Institute of Geophysics and Planetary Physics for five years prior to that.

Dr. Runnegar has a D.Sc. and Ph.D. from the University of Queensland, Australia.

Our interview follows.

**Suzan Mazur**: The NASA Astrobiology Institute established 10 years ago has been described by *Astrobiology* magazine – not the *Astrobiology* journal with links to NASA – as follows:

> "a virtual organization composed of NASA field centers, universities and research organizations that collaborate to study the origin of evolution, distribution and future of life in the universe. Astronomers, biologists, chemists, geologists, paleontologists, physicists and planetary scientists are involved with teams chosen via peer-reviewed proposals."

You were the director of the NAI a few years ago – roughly how many people are involved in NAI, and what would you say was your most important achievement as director?

**Bruce Runnegar**: *Astrobiology* is not a NASA journal or a NASA-influenced journal. It's a journal like any other journal in the community. It's run by an organization which is interested in making money.

The leader of that organization has chosen astrobiology as a developing field. Her business model is to find new fields of science and try to develop journals in those areas. There are a lot of NASA-connected connections with it, of course, because a lot of people working in astrobiology work with NASA or are funded by NASA. It's certainly not a NASA mouthpiece.

**Suzan Mazur**: You were director of NAI a few years ago?

**Bruce Runnegar**: Yes.

**Suzan Mazur**: How many people are involved with NAI and what would you say was your most important achievement as director?

**Bruce Runnegar**: When you ask how many people are involved, you've got to ask how many people are loosely connected. I'm not trying to avoid the question, but one way of looking at it is how many people are paid on a full time basis. And that's relatively few.

But many people are connected as academics, as advisors of postdocs and graduate students. People fully employed on NASA funding? There are a few hundred students and postdocs mainly. Academics get some salary assistance as is the tradition in American universities. NASA center members get more of their salaries from these sorts of sources. So a few hundred people are employed full time.

But there are probably, as you quoted in the article with Roger Buick, in the order of 1,000 people who are somehow connected with the NASA Astrobiology Institute. There are hundreds of institutions that are receiving some funding from the NAI.

There are 15 or 16 teams currently. They are also composed of consortia of other organizations, other universities, other research institutes. So if you total the number of organizations, it's in the order of hundreds.

My greatest achievement? What I was trying to do was to overcome the inherent difficulty of this kind of organization in that fierce competition to obtain membership through peer-reviewed processes. And then as soon as one is funded – and not just one person but tens of persons per team – then the idea is that these teams must then instantly learn to reverse that thinking and become collaborative and start to share resources, work toward a common good.

So taking the process somewhere down that track is what I was aiming for and regard as the most important achievement. Summarizing that, it's collaboration after competition.

**Suzan Mazur**: As former NAI director, can you tell me who makes policy at NAI and decides, for instance, that natural selection is the mechanism of evolution the public should know about?

**Bruce Runnegar**: No one decides that sort of thing in this organization. This is a scientific research organization. So what the public learns is what is believed to be the state of the art of science. The organization assists with that process trying to educate the public about new discoveries, but these discoveries like all science are subject to community acceptance and understanding.

Science, as you know, goes down the wrong track for many decades – as it did in the Earth Sciences when people didn't believe in continental drift and plate tectonics. Then eventually it takes a U-turn.

**Suzan Mazur**: The NAI supported publication, *Astrobiology Primer*. Were you heading NAI at the time this was supported?

**Bruce Runnegar**: Yes.

**Suzan Mazur**: The explanation for natural selection being promoted as the sole mechanism of evolution in that publication – the decision about that would have been whose?

**Bruce Runnegar**: **This is not a decision. The people who compiled that *Primer* are aiming to provide an educational service and so they write or wrote or acquired definitions of words. Basically trying to explain the jargon of astrobiology. Some of these explanations almost anybody could have different opinions about.** [emphasis added]

There were reviews of the document by scientists connected with the institute [NAI], but I don't think there's any attempt to provide a stamp of approval for any particular definition.

**Suzan Mazur**: I spoke with Lucas Mix, the editor of the *Primer*, about this. Mix is now writing a book on astrobiology, forthcoming from Harvard University Press. He referred me to the minor mention of neutral selection in the *Primer*, but natural selection was really the only mechanism of evolution discussed.

Do you think the next edition of the *Primer* will be updated to include current scientific thinking?

**Bruce Runnegar**: Natural selection is not a mechanism, it's the process by which the results of evolution are sorted.

**Suzan Mazur**: There are mechanisms which are being discussed in a major way, which were not covered in the *Astrobiology Primer*.

**Bruce Runnegar**: What do you mean by alternative mechanisms?

**Suzan Mazur**: Like self-organization for instance.

**Bruce Runnegar**: That's what I'm saying. There are a lot of mechanisms we know about. Self-organization being one of those kinds of mechanisms. Neutral evolution of genes. In other words, substitution without any change in the expression of the genes as far as we can tell. Those are sorts of mechanisms. But all of these are going on all the time producing a set of organisms which then natural selection can act upon.

You have to have variability in populations or in the biosphere. And that variability is produced by those mechanisms including self-organization. But ultimately it is competition or selection among those members of the biosphere that is the evolutionary process. That was Darwin's insight. Not the production of variation, but the ultimate effect of pruning by this natural selection process.

I think natural selection operates on all those mechanisms. That's, I guess, the point.

**Suzan Mazur**: There is a growing understanding that the peer-review process is rigged to maintain science status quo – that it holds back scientific progress.

Swedish cytogeneticist Antonio Lima-de-Faria calls this the "cycle of submission". He speaks of censorship in literature, like in the 2002 *Encyclopedia of Evolution*, which left out even A. Muntzing's work with *Galeopsis* in the 1930s where Muntzing crossed two different species and by doubling their chromosome number got *Galeopsis tetrahit* which occurred spontaneously in nature. Lima-de-Faria says no successive random mutations were needed.

In the case of *Astrobiology* journal's board there is a significant representation of NASA employees, which David Deamer recently acknowledged in an interview with me.

Do you see this as a problem, is there too much of a status quo operating in the peer-review process holding back discovery?

**Bruce Runnegar**: How is this any different from any other form of human activity? It happens in the arts. It happens in music. You remember the kerfluffle about the French Impressionists when they first came onto the Victorian scene. How dreadful their art was. It's just human nature to want to maintain the status quo and to keep things on the path one is accustomed to. And it's no different in science. People don't like to be shaken up with new discoveries. And new ideas do shake people up. So there is naturally some resistance. And it inevitably shows up in peer-review processes. It's just part of the way humans work.

**Suzan Mazur**: Are we any closer to answering the questions about the origin of life 10 years on?

**Bruce Runnegar**: I think we're getting a lot of new information that certainly bears on that question both in terms of how life actually works and that comes from the understanding we're getting from genomes. Not just the old idea that there is DNA that makes a gene and the gene makes proteins and the proteins all work together. And that all of the processes are much more complicated than people imagine. There are many more loops in the biochemistry of organisms. There are many cases where the RNA itself does the job and feeds back into the protein loop. So this whole system has become so much more complex. We understand the nature of life a lot more than we did 10 years ago. It's not just astrobiology. This is the advance of science as a whole.

**Suzan Mazur**: What are your thoughts about the origin of the gene?

**Bruce Runnegar**: Nearly every gene we see in nearly every organism is a modified copy of another gene that existed previously. A gene can stay more or less intact or evolve to have a new function by changing constituents of the gene, or a piece of a gene can be spliced with a piece of another gene in an evolutionary context.

As to how genes got started in the first place – that's a more complicated question. But we know from experiments in the last decade that you can actually make RNA that will evolve to do something in a lab in hours to days. So self-organizing of molecules that can do something worthwhile is certainly a very plausible hypothesis.

**Suzan Mazur**: Is the line between life and non-life arbitrary?

**Bruce Runnegar**: That's a more difficult philosophical question. That's something that we haven't really got a good idea of yet because we have such a diverse difference with the possible caveat of some viral-like particles. There it becomes rather difficult to separate. But that's all based on the same biochemical system. A bit like computer viruses are very separate from programs.

It's kind of a question that doesn't have any meaning if you don't have computers and codes in the first place. So it's all part of the living system and I don't think the definition matters in that case. With regard to life elsewhere, it's something we have no experience of, so it's hard to know whether we'd be more blurred with the inorganic world or not.

**Suzan Mazur**: Do we arrive at biological life via atomic, chemical, mineral evolutionary footprints?

**Bruce Runnegar**: That's what everybody thinks is the most reasonable.

**Suzan Mazur**: So everything is wired back to the atomic level?

**Bruce Runnegar**: First you have to build elements and you have to make them in stars. You know the story. Then you have to make compounds out of those elements. And if you go through and suggest what compounds are likely to be involved in life-like processes, you end up with a small short list. Carbon being one of those.

Then some of the fundamental things that we know about life – you don't want too many strong bonds amongst all these atoms because otherwise you end up with a solid that's like a piece of rock. You want some strong bonds and a lot of weak bonds.

There are obvious chemical reasons why life should prefer certain sorts of chemistry, etc. Yes, I think there is a hierarchy. Cells are part of that hierarchy, tissues, spatial organization, etc....

**Suzan Mazur**: Your current research is biotic and environmental events that accompanied the Cambrian explosion of multicellular organisms 500 million years ago. You're working with some of the oldest complex fossils in places like Namibia, Australia and elsewhere. Is it your view that it's natural selection every step of the way?

You do accept that mechanisms like self-organization come into play.

**Bruce Runnegar**: I said that before. The production of change, novelty has probably got many separate ways of doing that. Big jumps in things like going from one cell to multicellular organization may require different mechanisms, from changing the color of the stripes on zebras or something like that. But ultimately the choice of what survives to the present is competition among components of the biosphere that coexist. That's what we define by the words natural selection. It's the process that's operating on all these other mechanisms that makes natural selection kind of universal.

**Suzan Mazur**: Do you see any evidence of plasticity and a spontaneous emergence of body plans in development say 500 million to 600 million years ago?

**Bruce Runnegar**: Plasticity, again, I'm not sure what that means. Does that mean that one can transform into another as distinct from both diverging from a common kind?

**Suzan Mazur**: Ability to change form.

**Bruce Runnegar**: I think all that's nonsense. I'm a cladist. I believe that any two things have a common ancestor which looks like neither of them. If you take birds and crocodiles as an example – the common ancestor of any bird and any crocodile didn't look like either a bird or a crocodile. It looked like something else. Birds have evolved their characters from that common ancestor in one direction and crocodiles in the other direction.

So you can take the same logic back to that time. There was a common ancestor which didn't resemble anything that we see in the Cambrian. And it provided on one line of one branch something we would see and something on the other side we would see, but the common ancestor of those two were – we can't necessarily imagine what it looked like because we have no representative of it in the fossil record.

**Suzan Mazur**: How does the backbone form?

**Bruce Runnegar**: We know that flies that belong to a group of organisms known as arthropods have segments. The body is broken up into pieces that are all similar. Vertebrates have the same thing, better seen in a fish than a human perhaps. And we know from the genes that work those, that specify those segment patterns in vertebrates and the genes that specify those segment patterns in arthropods – that those genes predate the last common ancestor of those two groups.

If you go back from your average fruitfly and your average human and ask what did the common ancestor of these two animals look like – you have to say it didn't look like either a fly or a human. And then it becomes difficult to know because we don't have any clear representatives of that common ancestor. Probably because they were soft-bodied. Probably small. But we know that they were segmented.

Ultimately, the fact that bones are divided into vertebrae comes from that time when none of us had bones anyway. The bones came along later. But the patterning, the breaking of them up into pieces, comes from a genetic mechanism that existed long before bones.

So it's not amazing that we have a vertebral column. All the extra bits that go with it – the spine, the attached muscles – all those things can be explained in functional terms. The fact that we've got discs between the hard bones is probably just a functional thing because otherwise they'd grind each other away...

**Suzan Mazur**: You disagree with Roger Buick's analysis that the 3.47 billion year old rock in Australia presents evidence of the oldest metabolic process carried out by life – sulfate reduction by bacteria. He's been quoted in an NAI report as saying life from Mars probably colonized Earth.

You say the fractionated sulfur Buick found is due to exposure to hydrothermal vents and that the barite has always been barite.

Buick argues the fractionated sulfur he found has been through atmospheric processes. He told me in an interview that the debate has moved on substantially and that a French group published an important paper in *Nature* backing up his findings. He says your data has never been published outside conference abstracts.

Would you comment on why you think you're right and what that means in terms of the origin of life?

**Bruce Runnegar**: If you're talking about the original paper, forget about this business of going through the atmosphere because they only measured the ratio of two sulfur isotopes – the most common, which is sulfur 32. And the second most common, which is sulfur 34.

The convention is to take the ratio with the second most common 34, to 32 the most common. So you get a fraction, then you multiply by 1,000, because it's a small fraction.

If you look at just that in the modern world, you find sulfur-reducing bacteria like sulfate from the ocean, which on the scale that we're talking about, is about +20. And they reduce it to sulfide. That sulfide has a value which is negative. Let's not even worry about it. It's less than zero let's say.

So they've shifted the ratio from +20 down to something below zero. And the reason they do this is, or the way they do this is, because they prefer to use the lightest sulfur isotope – sulfur 32, than the slightly heavier sulfur 34, because it works better in chemistry.

Chemistry likes to do the same thing. But the reactions that chemistry does slow down those temperatures. So ordinary chemistry would do the same thing. It would reduce seawater sulfates to sulfides. But at the temperature of the modern ocean that process hardly works because without the assistance of life, no chemical reaction takes place.

In principle, the two processes don't have a different result. The only thing that makes it a signature for life is the temperature at which it happens. If it happens at ocean temperature, that's one thing. If it happens at a hot temperature, that's another thing. Then it could be either.

With the low temperature of the ocean, it's just not going to happen without the assistance of life. So then it comes back to what is the environment of the rocks in the deep distant past? Because you have to know what the temperature of this – what was going on at the time. What processes were going on when the sulfides were formed in these ancient rocks.

Roger started his career working in that part of the area. Did his Ph.D. on that sort of stuff. At that time they thought these sulfate minerals they found in the rocks had been produced by the evaporation of seawater. There was a sort of a marginal, at the edge of the sea lagoon where the water was evaporating. So it would have been seawater temperatures.

The arguments they used for saying these minerals were originally a different mineral are, I believe, not useful. If you accept the argument that they are now barium sulfate and they always were barium sulfate, then you have a different idea of what the original environment was like.

The most plausible thing is that they were made not at hydrothermal vents but at hydrothermal fluids. Because if you have hot rock under the ground, in this case a granite, then water coming from the surface goes down in cracks. It's fairly cool. When it gets down to near the granite it gets hot and so it rises. And these develop circulation systems in the rocks.

Somewhere it's going down and it's cool. Somewhere else it's hot. The vents occur where it comes up out of the surface. But I don't think these things were out of the surface. They were in the sediments and in the rocks when they were forming. This water going down would dissolve barium from the rocks. That's where the source of the barium is.

**Suzan Mazur**: When do you plan to publish?

**Bruce Runnegar**: I've got a year of sabbatical. So sometime this year. I'm hoping to get this into the press by the end of the year.

There are reasons why I don't think Roger is correct.

**Suzan Mazur**: Please continue.

**Bruce Runnegar**: I think these fluids that were coming up were carrying the barium. They were fairly hot. The barium reacted with the sulfate in the ocean which means that the deposition took place near the surface. Barium sulfate is very insoluable and that's how the sulfate deposits were formed.

Now the question of whether this sulfate was reduced by bacteria or not is another issue that's more complicated still. But I don't think the evidence is compelling. Let's put it that way.

**Suzan Mazur**: And what does this mean in terms of the origin of life?

**Bruce Runnegar**: I'm an astrobiologist, so I think it's important that we not raise false hopes. It doesn't matter very much on Earth. These kinds of investigations and experiments are relatively inexpensive on Earth. You can do all this for a few tens of thousands of dollars. But it's a highly different matter if you're going to Mars and spending a billion or two billion dollars returning samples or doing experiments in situ.

I think it's important that we be very confident about what we can say about data before we make these experiments elsewhere. That's why I'm taking the very cautious approach of wanting to be 100% sure before saying yes. Whereas I think some other people maybe are more enthusiastic, which is good for astrobiology perhaps, but perhaps not ultimately what we need for these kinds of investigations.

**Suzan Mazur**: Do you lean one way or another as to where life originated? Do you think it happened on Earth or perhaps on Mars?

**Bruce Runnegar**: We know that there are natural ways of transporting material between Earth and Mars we didn't know about 15 or so years ago. We also know from work done at CalTech that some of the rocks we've received from Mars, or at least one of the rocks that we've received from Mars, hasn't been heated enough to destroy life by the process of shifting it from Mars to Earth. So it's plausible that material containing microbes could have been transferred from Mars to Earth anytime in the last four billion years of Earth history.

So if life originated on Mars, it could conceivably have been transferred to Earth by this process. Perhaps less likely going the other way because Mars is a much smaller body and there's much less likelihood I think – I don't know this for sure – of getting materials going the other way. But that's the end of the story.

Nobody knows whether life has ever existed on Mars, and if it did, what kind of life it is. Whether it's the same as Earth life or not

**Suzan Mazur**: And how do you see life originating?

**Bruce Runnegar**: The origin of life one might expect would be a kind of set of experiments, but I think one of those experiments survived today. That's what natural selection does. Natural selection ultimately removed all the other failed experiments.

**Suzan Mazur**: And does natural selection exist throughout the Universe?

**Bruce Runnegar**: I think it should. It's a process that we see repeated in our own society. And that's the way businesses survive or fail depending on whether customers want the products. Right? That's natural selection in the case of business. It's a process that one would think would be universal.

**Suzan Mazur**: Do you think the Darwinian model is the model we should use in our economics, survival of the fittest?

**Bruce Runnegar**: I'm not an economist. And that's an expression which raises all sorts of emotional responses in the audience. But you may know that some of the most interesting models of evolution involve things like game theory. The famous game called the Prisoner's Dilemma where altruism is actually one of the most important components of evolutionary systems. So natural selection doesn't necessarily mean cutthroat competition, and that's what businesses I think also realize – that it's sometimes better to promote biodiversity of businesses rather than than try and grind the opposition to a close.

I'm no expert on this, but you only have to look at some of the ways businesses work collectively to realize that it's not as black and white as much as this description you just said [survival of the fittest] might indicate. I'm all in favor of it in principle, but I'm not in favor of a caricature of the process.

**Suzan Mazur**: Do you recognize the new Extended Synthesis – the reformulation of the neo-Darwinian theory of natural selection that was kicked off at Altenberg in July?

**Bruce Runnegar**: I know nothing about it I'm afraid. I've been busy in the field.

**Suzan Mazur**: What is your connection currently to NASA?

**Bruce Runnegar**: I'm two years beyond some of the post-employment restrictions. I'm firmly disconnected from NASA for the last two years.

**Suzan Mazur**: Is there a final comment you'd like to make?

**Bruce Runnegar**: I've spoken at length because sometimes a brief comment leads to misunderstanding. I want it to be clear where I stand on these issues because they are contentious....

# NASA Humanist Chris McKay: Where Darwinism Fails

*September 16, 2008*
*4:12 pm NZ*

Over the phone I detect a touch of William Shatner's Kirk in the voice of NASA astrobiologist Christopher P. McKay. McKay admits he was inspired by the television series *Star Trek* 30 years ago and the "great voyages of discovery". But while most of his professional life has been at NASA Ames Research Center in the Space Science Division, beginning as a Planetary Biology Summer Intern in 1980, he objects to being typecast saying, "I'm a scientist and not a humanist? That's idiotic."

McKay is now a planetary scientist at Ames researching the evolution of the solar system as well as the origin of life. He's been involved in planning Mars missions including the 2009 Mars Science Lander. He's an authority on Titan (Saturn's moon) and was co-investigator on the Titan Huygen 2005 probe. McKay is also Program Scientist for NASA's Robotic Lunar Exploration Program.

He says he does his best thinking in extreme Mars-type environments – the Arctic, Antarctic, Siberia and Chilean desert. In 1994, The Planetary Society honored him with the Thomas O. Paine Memorial Award for the Advancement of Human Exploration of Mars.

McKay now serves on the board of directors of The Planetary Society and on the editorial boards of *Astrobiology* journal as well as *Planetary and Space Science* journal. He studied physics and astrophysics as an undergraduate and has a Ph.D. in AstroGeophysics from the University of Colorado.

Chris McKay is author/editor of several books, among these: *Case for Mars II, Comets and the Origin and Evolution of Life, Earth's Climate, From Antarctica to Outer Space: Life in Isolation and Confinement*.

My telephone interview with Chris McKay follows.

**Suzan Mazur**: You were co-investigator for NASA's Phoenix Mars mission in May. We were told that perchlorate was found in the soil. And that clay and possibly methane exist on Mars as well. What do each of these signal?

**Chris McKay**: Let's start with the methane. This was evidence from ground-based telescopic spectroscopy and the European Mars Express mission. There's a lot of controversy about the methane and there's still not yet a coherent story about the data and its interpretation. I actually belong to the camp that believes it's not real, that the methane data is a mistake.

**Suzan Mazur**: What would the existence of methane signal?

**Chris McKay**: If the observations are valid and there is methane on Mars, it tells us that there is a very strong source of methane and that it varies on short time scales, much less than a year. Biology is one possible cause.

**Suzan Mazur**: What about the clay?

**Chris McKay**: Clay is much more clear. There's evidence from orbital data for phylosilicates, which is a fancy word for clay. That's pretty solid and not surprising. It's consistent with what we understand about Mars. And the distribution of the clays is interesting.

**Suzan Mazur**: What is interesting about the distribution of clays?

**Chris McKay**: The clay is found mostly in the ancient regions of Mars, probably indicating that these are the locations which had water.

**Suzan Mazur**: This is clay that has fossilized?

**Chris McKay**: Not fossilized but old.

**Suzan Mazur**: And that means?

**Chris McKay**: It's a relic from some time when Mars had a lot more water. This is the leftover mud from an early wet muddy period.

**Suzan Mazur**: What is the significance of finding perchlorate?

**Chris McKay**: The perchlorate recently detected by the Phoenix mission adds to the mystery of the Martian soil. Perchlorate is an oxidizing form of chlorine and we do find it in deserts on Earth, for example the Atacama desert in Chile. Perchlorates are not bad for life, but they are not good for life either. They certainly do not rule out biology.

**Suzan Mazur**: Aside from drilling down a kilometer or two into Martian soil to identify what is there, which some say could take anywhere from decades to centuries – you're keen on the idea of terraforming. Nudging Mars back to its previous life, if Mars had one.

You refer to life on Mars as a possible "second genesis". Life on Earth as the "first genesis"? Would you say a bit more about your vision?

**Chris McKay**: There are two thoughts here. One is the notion of a second genesis: Did Mars have life and was that life a second genesis. Was there a separate, independent origin of life on Mars? Did that occur?

That's a question about the history of Mars that we seek to answer with robotic missions or with human missions. It's a science question.

If we find there was life on Mars and that it represents an independent origin of life – what I call a second genesis of life – that's wonderfully important scientifically as well as philosophically. Scientifically it gives us the opportunity for the first time to compare two types of life. All life on Earth is one type. If we find a second genesis on Mars, we will then have the opportunity to compare biochemistries for the first time.

It's also wonderful philosophically because if right here on our own solar system life started twice, once on Earth and once on Mars, then it's clear the Universe is full of life. That it starts on all Earth-like planets.

As long as we only have one example of life (Earth life), we never really know if it isn't just a cosmic fluke. So to my mind the search for a second genesis is the most important science question about Mars. That is the science-driver in terms of understanding Mars' past history and potential for life.

Now there's another question about Mars' future. The second genesis is the story about Mars' past. The story about Mars' future is: Could Mars have life in the future?

That would involve global change. We'd have to warm Mars up once again to make it a planet suitable for life.

**Suzan Mazur**: Critics of terraforming ask how we would do this on Mars when we have big environmental problems on our own planet. Then there's the cost of terraforming Mars to consider as well as the dangers of messing around with the Martian environment without fully knowing what may be out there. Would you comment?

**Chris McKay**: There are two concerns. One is the cost. And I think the cost is probably small. We're not talking about a massive engineering project. We're actually talking about producing gases on Mars the way we produce them on Earth and letting nature take its course.

The normal scenario people have when they think about terraforming is some huge science fiction-like enterprise –

massive fleets of spacecraft shining giant lasers down on Mars. That's not what I think would make sense.

What would make sense would be producing gases at a low level on Mars – the same gases we're producing on Earth – supergreenhouse gases, which are now warming up the Earth. Then once Mars warmed up, letting life follow its own evolutionary history. The actual effort involved would be quite modest.

But there's an issue separate from the economic issue, which is: Is this something we want to do and how does it relate to Earth?

Sometimes people say we need to do this so that we have a lifeboat after we've messed up the Earth. So we'll have somewhere to go. That is not a practical option and it's ethically absurd.

**Suzan Mazur**: Anything more you'd like to say along those lines?

**Chris McKay**: Yes. If Mars had a separate origin of life, there's a possibility life is still there on Mars, even if in a dormant state. In that case I have advocated bringing Mars to life, back to Martian life. "Mars-aforming" rather than terraforming.

**Suzan Mazur**: I visited Saudi Arabia a few times and have seen the Saudi successes at reclaiming the desert.

**Chris McKay**: We're also reintroducing the wolf here in Yellowstone Park. Restoration ecology is a watchword for this millennium.

It used to be ecology meant doing nothing. Ceasing to do bad things represented ecological action. Now we realize doing positive things is important too. Civilization has become ecologically pro-active. Reintroducing the wolves in Yellowstone is an example.

We could bring a planet back to life. It would be the first really positive thing on our collective resume of space missions.

**Suzan Mazur**: What about space law? In 1967 we had the Outer Space Treaty, which said there would be no nukes or any other WMDs in orbit of Earth or installed on the Moon or any other celestial body. Also setting out that space is *res communis* – no one country has jurisdiction to make laws in space. 98 countries signed and 27 ratified. Did the US and other major powers ratify this treaty?

**Chris McKay**: I think so, but I don't really know that.

**Suzan Mazur**: Then in 1979 we had the Moon Treaty based on the UN Convention of the Laws of the Sea, which spelled out that the profits of space would be shared equally among nations. Leading the charge that resulted in the US not ratifying or even signing the Moon Treaty was the L5 Society who were promoting Gerard O'Neill's ideas of space habitats (the L5 women said they would not become pregnant until they were in the space colony). L5 were also advocates of terraforming. 13 countries did ratify the Moon Treaty but no major powers.

Those ratifying included: Australia, Austria, Belgium, Kazakhstan, Lebanon, Mexico, Morocco, Netherlands, Pakistan, Peru, Philippines and Uruguay. France and India signed, but Russia, China and the US neither ratified nor signed.

I was writing stories for *Omni* magazine at the time and asked Malcolm Forbes what he thought of the Moon Treaty. He was not in favor of the Moon Treaty. Here's some of what he told me, which he reprinted in *Forbes*:

> "I think it's a nice academic theory, but the point is, who's going to spend all the money to dig out the ore if all of it has to be turned over to the commune of nations? You obviously can't go out

> and stick a flag down, as in the old colonial days, and say the moon is yours. This is your Saturn. But you just can't remove incentive and say everything belongs to everybody. That would mean nothing belongs to anybody, and nobody would then go get it...
>
> I think that if the French find a lot of ore on a particular asteroid, then France should be able to sell the ore in the world marketplace. Then somebody else will go after another asteroid for ore, as entrepreneurs have done on Earth for oil and for everything else. Competition makes people go seek it, and mankind profits universally. Though a drug company may own the rights to a certain medicine, mankind globally eradicates a certain disease."

Does current space law address this issue about sharing the profits of space and what are your thoughts?

**Chris McKay**: I have thoughts on it but I'm not involved in space law at all. I think these points are irrelevant.

**Suzan Mazur**: Really?

**Chris McKay**: They are based on what I think is a false assumption, which is that there is money to be made in space. I think that's smoke & mirrors. It's the same smoke & mirrors that they applied to the space station and the production of wonder drugs in microgravity.

**Suzan Mazur**: So where are we now then in the discussion about the commercialization of space? In 1985 *Aviation Week* launched a magazine called *Commercial Space: The First Business Magazine of the Space Age*. The thinking was that space was "not just adventure, exploration and national prestige anymore" – in fact, 2,100 companies were supposedly supporting space activities at the time.

I noticed that *Aviation Week* has a web page article called "International Commercial Space Development for the Future 50 Years", but the original *AW* magazine *Commercial Space* seems to be defunct.

Also, I remember having a lengthy conversation with someone at Shearson Lehman in the mid 80s about investments in space. You don't think the thunder's still there regarding business in space?

**Chris McKay**: I meant the way *Forbes* was describing it. Mining in space. There's business in space, of course. The business in space is Earth observation and tourism, the only businesses in my view that make sense at all in space.

**Suzan Mazur**: Earth observation.

**Chris McKay**: In my view, satellites in orbit looking at Earth is not really space business. It technically is in industry parlance.

**Suzan Mazur**: So these 2,100 companies that were supporting space activities at the time?

**Chris McKay**: I bet 99.999% of them were supporting launch and Earth observation, Earth satellites, Earth communications.

**Suzan Mazur**: You don't think the drug companies are sending up experiments.

**Chris McKay**: There have been no drugs developed in space. There have been no companies beating down the door. The only business in space has been the use of space as a platform to look at or talk to Earth. That's a huge industry.

When I go out into the field, I have a little dish and I can get Internet through satellite. I love it! I get on Google Earth and I can look into my backyard and love it. That's not space industry in my view, that's Earth industry. That's not what I mean by space industry.

What Forbes was implying was that there's gold on an asteroid and someone's going to go out and mine it and sell it on Earth. That's what I say is bogus.

**Suzan Mazur**: That's science fiction.

**Chris McKay**: It's not just science fiction, its absurd. Science fiction when it's done well is about things that we can't do but we can think possible. Absurd are things that just make no sense at all.

**Suzan Mazur**: Have the number of space interest groups been expanding or has the pouf gone out of the soufflé in recent years about interest in space?

**Chris McKay**: I think they're high right now, partly because NASA is back on the track to go to the Moon and Mars. But there's still – you go to NASA headquarters and there are still people talking about how we're going to make money mining oxygen on the Moon and we're going to have return on investment. We're going to sell oxygen mined on the Moon. I think it's so stupid that I've stopped even arguing about it.

I don't think there is a mineral-based economic activity on the Moon or Mars or asteroid belt that's going to pull private industry. But people still believe there is.

The L-5 Society/Forbes view that you just quoted – which is the view that there are mineral resources in space and we're going to get rich by mining asteroids and the Moon, selling those minerals to passing spaceships or bringing it back to Earth – I just think that's absurd.

Tourism is not absurd. Clearly there is a lot of interest and a lot of money to be made in tourism. Serious contenders are putting money forward in space tourism, like Virgin Galactic. Nobody is spending real money developing a mining operation on the Moon.

**Suzan Mazur**: Now I understand the Astrobiology Program funds are hurting. Is that right?

**Chris McKay**: That's another matter. The science will continue at some small level. Science funds will go up and down as fashions dictate. Sometimes it'll be geology and astrobiology. They'll always be in there. But it's always going to be a low level government-sponsored activity.

**Suzan Mazur**: What of the Astrobiology Program?

**Chris McKay**: Any science activity. Lunar geology, Martian astrobiology, the search for life, understanding volcanoes on the Moon and Mars – all of these are worthwhile science activities and they will all occur, but will be a low-level activity.

**Suzan Mazur**: Seems peculiar with people so interested in the origin of life – I would think there would be more resources put into this.

On the theory side of origin of life – are you involved in the review of papers and would you tell me what you look for in accepting or rejecting a paper? What reservations do you have in reviewing a paper coming from outside your peer circle?

**Chris McKay**: Mostly, is the paper presenting new ideas or new data. A lot of astrobiology papers written on the origin of life are just somebody's afternoon speculation.

**Suzan Mazur**: You're on the board of...

**Chris McKay**: I'm on the editorial board of several journals.

**Suzan Mazur**: Which ones?

**Chris McKay**: *Astrobiology, Planetary Space Science,* probably some others I can't keep track of. I get a lot of papers to review and I spend a lot of time reviewing papers. I think I'm the favorite reviewer for anything that's astrobiologically speculative.

A lot of people lately – because astrobiology has become so fashionable – like to write what you could call Sunday afternoon theories. Well, maybe life could be based on boron, for example, and they write about it – all just speculation.

I insist that papers have new ideas, important new ideas or important new data. That they don't just be somebody's random ideas like wouldn't it be interesting if there was life on a planet that was really hot and based on heavy metals and...

**Suzan Mazur**: So a paper needs math.

**Chris McKay**: Not necessarily math, it needs substance. It needs to have some meat to it. I'm using the word in the old English sense. A paper can have an important new idea or important new data and develop that idea or data. What I'm coming across is a lot of papers that are basically sophomore term papers on the origin of life. They review the field and then speculate about possibilities.

Even though the paper's well written, I've read this 50 times before. If this were a sophomore term paper in an astrobiology class, maybe I'd give it an A. But it doesn't need to be published in the literature because it doesn't add anything to our collective understanding.

**Suzan Mazur**: What percentage of papers do you actually give a nod to?

**Chris McKay**: Oh I'm pretty generous – about 80% or 90%.

**Suzan Mazur**: Really.

**Chris McKay**: Yes. I'm known as the soft reviewer. That's just the way I am. My view of reviewing is that it's not the onus of the reviewers to be the gatekeepers. My name doesn't get attached to the paper. It's not a problem for me if a paper gets published that's garbage. The author's name is attached to it.

The primary responsibility for quality and acceptability is the author's.

I reject only in egregious cases, where I think it would be a real waste of the journal, in the sense that I'm a guardian of the ink space of the journal. We have finite pages we can publish each year and I don't want to waste those pages.

Again, I don't assume intellectual responsibility for the papers. The authors do that. I give authors broad benefit of the doubt.

**Suzan Mazur**: How many of these papers coming in for review would you say you're reading on a weekly basis?

**Chris McKay**: I'd say it's at least one a week. It comes in waves. Right now I have so many that I have to keep a separate folder on my computer called papers and proposals to review.

**Suzan Mazur**: Are you saying you're overwhelmed with papers at this point?

**Chris McKay**: One a week is overwhelming. When astrobiology was less popular there'd be two or three a year. But one a week is overwhelming.

**Suzan Mazur**: It's interesting that you give the nod to 80% or 90% of papers.

**Chris McKay**: Most scientists submit a paper that does have new ideas and new data. Scientists realize their name is on the paper. So the papers are generally not just random ruminations.

**Suzan Mazur**: These papers don't require data, but it helps to have data. The important thing is to have a clear idea and develop it.

**Chris McKay**: If you're publishing a paper without data, then you've got to have some pretty good new ideas. There are papers which present new ideas and explore the implications and present testable hypotheses derived from those ideas. That's an interesting paper. Let's publish that.

**Suzan Mazur**: What is your perspective on the line between life and non-life being arbitrary?

**Chris McKay**: I don't think we understand that division. It's useful to think of it in two ways. One is that the division is sharp. That it's like a phase transition. Like the difference between water and ice. It's rather sharp. And then it's meaningful to talk about living systems as distinct from non-living systems or pre-biotic and biotic.

The other view is that it's not sharp. That it's gradual. That it's a continuum of a state. And then it becomes hard to define when something is alive and when it is not.

**Suzan Mazur**: That's when it's arbitrary.

**Chris McKay**: We as scientists don't know.

**Suzan Mazur**: But what is your best guess?

**Chris McKay**: My best guess is that it's a sharp transition. I guess that based on the nature of life which is this feedback between information and matter. You have information stored in the genome and then you have material systems which are organic molecules which implement that. The feedback is self-amplified feedback. Typically self-amplified feedbacks have sharp ons and offs.

**Suzan Mazur**: Now you say in searching for life elsewhere in the Universe, the logical things to look for are energy, carbon, liquid water, nitrogen, sulfur and phosphorus. But you don't say look for natural selection. Do you think Darwinian natural selection exists throughout the Universe?

**Chris McKay**: Yes. Natural selection, I think, is the essential aspect of life no matter where we find it. It would be great to have some way to detect natural selection, but we're unlikely to be able to. We have a hard time detecting it here on Earth and showing that it's occurring.

But we see the products of it. In our biochemistry we see the results of billions of years of natural selection and optimization. So we detect natural selection indirectly.

If we were to go to Mars and find a dead rabbit on the surface of Mars, a Martian rabbit, that is neither alive nor is it undergoing reproduction and selection. But it would be proof of Darwinian selection on Mars because that rabbit or the Martian animal, the dead Martian animal would be proof of natural selection because only natural selection could produce that dead animal. And so we would have found evidence of life, evidence of natural selection – and if it was an alien biochemistry, evidence of a second origin of life, all from having found a single dead animal.

There's an unfortunate problem in the English language where life refers to a process, refers to an individual, and it refers to a collective phenomenon that has a history. When people talk about defining life – they mix all those together in a messy way and it makes no sense...

**Suzan Mazur**: What about other mechanisms of evolution, self-organization and self-assembly? They precede natural selection?

**Chris McKay**: Something had to precede Darwinian natural selection. **The Darwinian paradigm breaks down in two obvious ways.**

**First, and most clear, Darwinian selection cannot be responsible for the origin of life. Secondly, there is some thought that Darwinian selection cannot fully explain the rise of complexity at the molecular level.** [emphasis added]

**Suzan Mazur**: So you're saying Darwinian natural selection sets in at what point?

**Chris McKay**: I think it must set in after life has started. After there's a genome, genotype. That's the one obvious place where Darwinian natural selection fails – is in the origin of life. It can't be Darwinian all the way down.

**Suzan Mazur**: At what point did the gene set in?

**Chris McKay**: We don't know. That's the question. It's got to do with whether the transition to life is abrupt or gradual.

**Suzan Mazur**: What is the gene?

**Chris McKay**: The gene in a general sense is anything that stores information in an algorithmic way. Stores instructions how to build something. It doesn't have to be DNA, it could be RNA or it could be something else. But at some point life invented software.

I think the language of computers is very useful here. There's a distinction between hardware and software. Darwinian selection only works when there's software. And everything that's prebiotic is hardware.

At some point life got onto software. And that's when Darwinian selection could begin. Darwinian selection can't work on hardware by definition because Darwinian selection involves inheritable traits. Only a system that has software has inheritable, mutable traits. It doesn't have to be DNA, but it has to be software. And it has to record algorithmic information, instructions.

**Suzan Mazur**: You began your association with NASA in 1980 as a Planetary Biology Summer Intern at NASA Ames – which is where you are today.

**Chris McKay**: Yes. Still here. Same building.

**Suzan Mazur**: In a different capacity though. Those *Star Trek* episodes about a positive future really meant something.

**Chris McKay**: Yep.

**Suzan Mazur**: We've gone through a dark period in recent years – could more films about a positive future actually help bring one about do you think? How close is science fiction and science reality?

**Chris McKay**: Let me not just talk about films. Let me talk about literature and the humanities. The humanities is the study of human things in human terms. And they're incredibly important to defining our human view. Science doesn't tell us how to live our lives or how to strive for a better world. The humanities do that. And so there's a terribly important role for literature, film, art.

Science gives us a lot of valuable information. But it's the humanities that tells us about the human condition and what motivates us to make the world a better place – literature and art. It's our understanding of our humanness in human terms. That's what the humanities are.

I am a student of the humanities because I am a human being. I completely reject a distinction between science and the humanities. I'm a scientist and not a humanist? That's idiotic.

**Suzan Mazur**: Science and art have a close connection historically. Thinkers like da Vinci, for instance, and so many other examples.

**Chris McKay**: Even in the non-genius end, as a human being and part of this culture I interact with this culture. I find motivation and inspiration and learn from being part of this culture. The human aspect of my environment is very important to me.

Literature, science fiction, are part of how we together create an image of the world as we would like it to be. So when we write stories about a positive future, we are in a very real way telling ourselves we shall make it so. We shall make a positive future. I'm a big supporter of that. It's an important part of my own motivation.

# Chapter 15
# The Rome Abstracts: "Evolutionary Mechanisms"

*February 14, 2009*
*11:44 am NZ*

Why have the two major evolution conferences of the year (this and last) been hosted outside the United States? In July 2008, we had Altenberg, Austria and the "Extended Synthesis", and on March 3-7, 2009 in Rome, "Biological Facts and Theories: A Critical Appraisal 150 Years After *The Origin of Species*" under the patronage of the Pontifical Council for Culture.

The March conference will reconsider "issues surrounding evolutionary biology" and topics such as evolutionary mechanisms (a full day of talks), human origins, and the philosophical and theological aspects of evolution, with dozens of the most substantial evolutionary thinkers on Earth present. It is a collaboration of the University of Notre Dame and the Pontifical Gregorian University. But why couldn't such an event take place in Indiana?

And while there has been some discussion about streaming the meeting live on the Internet, PGU organizers have advised that the conference abstracts presented to the media this week were in Italian, meaning the English-speaking world may not be getting the message of just how central evolutionary mechanisms are to this highly important conference. Nor that scientists who will speak about evolutionary mechanisms think Darwinian natural selection should be relegated to a significantly lesser role.

So I asked the following scientists who are presenting in the "Evolutionary Mechanisms" sessions if they'd share their abstracts – in English. They are: **Stuart Newman** – New York Medical College; **Scott Gilbert** (Rome conference scientific committee) – Swarthmore College; **Stuart Kauffman** (Rome conference scientific committee) – University of Calgary and Harvard University; **Lynn Margulis** – University of Massachusetts-Amherst and Oxford University; **Robert Ulanowicz** – University of Maryland; **Jeffrey Feder** – University of Notre Dame; **Francisco Ayala** – University of California-Irvine; **Jean Gayon** – Universite Paris 1. Their abstracts follow. [Note: A week or so after this story appeared, PGU made the decision to release all conference abstracts in English.]

**STUART NEWMAN**

> This conference is one of the few worldwide in the anniversary year of Darwin and *The Origin of Species* that is providing a forum, as well, for concepts of evolutionary change other than natural selection of incremental variations. Over the past two decades our understanding of the dynamic and plastic nature of developmental systems has provided insight into major modes of evolutionary innovation left out of the well known Modern Synthesis. It is to the credit of the conference organizers that in addition to the legacy of Darwin, their scientific program will include this newer story.
>
> – **Stuart A. Newman**, Professor of Cell Biology and Anatomy, New York Medical College

## A "Pattern Language" for Evolution and Development of Animal Form

Ancient animals arose from unicellular organisms that had billions of years of genetic evolution behind them. In this talk I will consider the role played by a core set of "dynamical patterning modules" (DPMs) in the origination, development and evolution of multicellular animals, the Metazoa. These functional modules consist of the products of certain genes of the "developmental-genetic toolkit," in association with physical processes and effects characteristic of chemically and mechanically excitable systems of the "mesoscale" (*i.e.*, linear dimension ~0.1-10 mm).

Once cellular life achieved this spatial scale by the most basic DPM, cell-cell adhesion, a variety of physical forces and effects came into play, including cohesion, viscoelasticity, diffusion, spatiotemporal heterogeneity based on lateral inhibition, and global synchronization of oscillatory dynamics.

I will show how toolkit gene products and pathways that pre-existed the Metazoa acquired novel morphogenetic functions simply by virtue of the change in scale and context inherent to multicellularity. I will show that DPMs, acting singly and in combination with each other, constitute a "pattern language" capable of generating all metazoan body plans and organ forms.

Another implication is that the multicellular organisms of the late Precambrian-early Cambrian were phenotypically highly plastic, capable of fluently exploring morphospace in a fashion relatively decoupled from both function-based selection and genotypic change. The stable developmental trajectories ("developmental programs") and morphological phenotypes of modern organisms would then have arisen by stabilizing selection that routinized the generation of morphological motifs that were originally manifestations of the physical properties of multicellular aggregates.

This perspective provides a resolution of the apparent "molecular homology-analogy paradox," whereby widely divergent modern animal types utilize the same molecular toolkit during development. It does so however, by inverting the neo-Darwinian tenet that phenotypic disparity was generated over long periods of time in concert with, and in proportion to, genotypic change.

**SCOTT GILBERT**

> The significance of evolutionary mechanisms being central to the Rome evolution conference? Including the notion of evolutionary mechanisms gets around being tied down to one paradigm. A very wide net has been cast even within the Evolutionary Mechanisms discussion. There are some things not being discussed. And other things that are.
>
> Lynn Margulis is talking about symbiosis and the origin of species – a model of cooperation as opposed to a model of strict competition – and Stuart Kauffman and Stuart Newman are addressing the physical antecedents and nongenetic evolution. These are things that are not in the normal paradigm of competitive natural selection.
>
> Natural selection will be represented – Doug Futuyma is speaking. People will be discussing natural selection. But in a way natural selection has been proven so often and so definitively, it's like discussing – are skeletons made of bones?
>
> Natural selection occurs. Natural selection occurs within species. But natural selection alone cannot

explain how butterflies got their wings. How the turtle got its shell. Once you have variation within species, then natural selection can work. Do I think natural selection should be relegated to a less important role in the discussion of evolution? Yes I do.

–**Scott F. Gilbert**, Howard A. Schneiderman Professor of Biology, Swarthmore College

**Evolutionary Developmental Biology: Evolution by Epigenesis**

In 1893, Thomas Huxley, wrote, "Evolution is not a speculation but a fact; and it takes place by epigenesis." Note that evolution's chief defender did not complete his sentence with the phrase "natural selection," for Huxley was interested in the generation of the diversity needed for natural selection. That phase of evolution was regulated by development. Recent work has established five main mechanisms for the generation of anatomical diversity through changes in development, and this talk will review them and provide examples from the recent literature.

These mechanisms are:

(1) **Heterochrony** (changing the time or duration of developmental phenomena or gene expression)

(2) **Heterotopy** (changing the placement of developmental phenomena or the cell types in which a gene is expressed)

(3) **Heterometry** (changing the amount of gene expression in a manner sufficient to alter the phenotype)

(4) **Heterotypy** (changing the sequence of the gene being expressed during development)

(5) **Heterocyberny** (change in the "governance" of a trait from being environmentally induced to being genetically fixed)

These mechanisms have profound significance for how new traits can be generated, how they become integrated into developing organisms, and how they can become propagated through a population. It is argued that adding these developmental data and contexts provides a new and more complete theory of evolution, including a theory of body construction along with a theory of change.

**STUART KAUFFMAN**

The fact that the Pontifical Gregorian University and STOQ are hosting a meeting on evolution in the Darwin year is of major significance culturally. In the United States, only a few years ago, Intelligent Design proponents attempted yet again to ram a poverty struck version of science down the throats of American school children. Yet even one of its leaders, Michael Behe, admitted in court that astrology would count as a science under their definition of science.

At present, some members of the U.S. Congress, and George Bush, to appease the religious right, support Intelligent Design, whose statistical arguments are bogus largely through their failure to understand the concept of Darwinian "exaptations" in the emergence of novel functions in evolution. Thus, this stance by a Church that speaks for a billion Catholics is honest, timely and brave.

Darwin changed our thinking as much as any mind has. The ramifications from Darwin are still being unpacked, including a possibility that may have deep spiritual importance: no natural law may suffice to describe the full evolution of the biosphere, human economy, and human culture.

**In its place, in my view, is a ceaseless creativity in which we cannot know before hand even what CAN happen, let alone what will happen. But then reason, the highest human virtue of the Enlightenment, is an insufficient guide to live our lives. We need reason, emotion, intuition, imagination, story, metaphor and more. In fact, we need a new Enlightenment.** In my own view, we can consider this fully natural creativity in the universe God. We are its children.

**– Stuart Kauffman FRSC, MacArthur Fellow**

### Are Cells Dynamically Critical

The genetic regulatory network in humans has some 23,000 genes, among which are at least 2,040 transcription factor genes. These TFs regulate one another's transcriptional activity and those of genes that are regulated but not regulating. Work on yeast gene networks shows that they appear to be one large interconnected network. Probably the same is true for humans This genetic regulatory network is a non-linear dynamical system of high complexity. Modeling genes as binary, on, off, devices and studying large "random Boolean networks" has shown that these networks, and piecewise linear networks, and linear ordinary differential equation networks all show the same generic behaviors: They have an ordered regime and a chaotic regime, separated by a dynamically "critical" phase transition. Such networks also show dynamical attractors, somewhat analogous to lakes in a mountainous region, each fed by streams in its drainage basin.

The two central hypotheses I will explore are that cell types correspond to dynamical attractors, the lakes in the analogy, and cell differentiation consists of signals or noise that lifts the system from one lake into a different basin of attraction, or drainage basin. Critically, I suppose, with my colleague Dr. Sui Huang, that cancer cells as also attractors: the Cancer Attractor Hypothesis, and will discuss current high throughput high content screening of small molecule chemical libraries that are able to induce differentiation of cancer cell lines into non-proliferating mature cells. We hope this will afford a systematic approach to cancer "differentiation therapy". The second hypothesis I will discuss is that evolution has selected cells that are dynamically critical. Importantly, such systems optimize information transfer, and information storage. In addition, we have initial results that critical networks also maximize power efficiency – a sensible target of natural selection. Even more importantly, I will discuss recent work suggesting that real cells are actually critical, and that cell types are, indeed, dynamical attractors.

**LYNN MARGULIS**

**Origin of Evolutionary Novelty by Symbiogenesis**

Whereas speciation by accumulation of "random DNA mutations" has never been adequately documented, a plethora of high-quality scientific studies has unequivocally shown symbiogenesis to be at the basis of the origin of species and more inclusive taxa.

Members of at least two prokaryotic domains (a sulfidogenic archaebacterium, a sulfide-oxidizing motile eubacterium) merged in the origin of the first nucleated organisms to form the earliest eukaryotic organism in the mid-Proterozoic Eon (c. 1200 million years ago.)

Such a heterotrophic, phagocytotic motile protoctist was ancestral to all subsequent eukaryotes (*e.g.*, other protoctists, animals, fungi and plants).

The defining seme of eukaryosis, the membrane-bounded nucleus as a component of the karyomastigont, evolved as *Thermoplasma*-like archaebacteria and *Perfilievia-Spirochaeta*-like eubacteria symbiogenetically formed the amitochondriate LECA (the last eukaryotic common ancestor). Their co-descendants (that still thrive in organic-rich anoxic habitats) are amenable to study so that videos of them can be shown here.

There are no missing links in our scenario. Contemporary photosynthetic (green) animals (*e.g., Elysia viridis, Convoluta roscoffensis*), nitrogen-fixing fungi (*Geosiphon pyriforme*), cellulose digesting animals (cows, *Mastotermes darwiniensis* termites) and plants (*Gunnera manicata*) make us virtually certain that Boris Michailovich Kozo-Polyansky's (1890-1957) analysis *Symbiogenesis: A New Principle of Evolution* (1924) was and still is correct

Symbiogenesis accounts for the origin of hereditary variation that is maintained and perpetuated by Charles Darwin's natural selective limitations to reaching the omnipresent biotic potential characteristics of any species.

**ROBERT ULANOWICZ**

**Process and Ontological Priorities in Evolution**

Charles Darwin, a fervid admirer of Isaac Newton, nonetheless described evolution as a process, rather than as the action of laws upon objects. Against this bold initiative, the "Grand Synthesis" of Fisher and Wright and the ensuing discoveries in molecular biology ushered in the Neo-Darwinian scenario wherein ontological emphasis has reverted to material objects and mechanisms. Other life sciences, however, continue to lend themselves more naturally to description in terms of processes.

The dynamics of ecosystems, for example, can be seen to rest upon a set of fundamental postulates corresponding to the attributes of processes. Mutuality stands at the ontological core of this perspective, known as "process ecology". By comparison, competition is seen to be accidental and derivative. Unlike in the Newtonian/Darwinian schema, selection in process ecology can occur internal to the system, rather than solely via the exogenous agency of "natural selection".

The monist dictum of "survival of the fittest" appears to relate to only one side of a broader Heraclitean/Hegelian agonism. Such discrepancies with orthodox evolutionary theory suggest that a far richer picture of evolution (and the ethos that it informs) may be possible by reverting to Darwin's initial instinct to describe living nature primarily as process. Adopting the process perspective mitigates many of the ostensible conflicts between science and religion.

**JEFFREY FEDER**

**The Mystery of Speciation**

Charles Darwin once described speciation as that "mystery of mysteries". However, this is one mystery that is now arguably solved In this talk, I will discuss advances in our understanding of how new species form since Darwin first posed the problem. I will outline current views regarding the geographic context and genetic bases for speciation. I will also examine how and why barriers to gene flow evolve - the crux of the speciation problem - highlighting a few case studies demonstrating incipient speciation in action I will conclude by discussing current research directions in the field. Although we may understand the general mechanisms generating new species, much remains to be learned. In particular, we are entering an exciting new period of synthesis in which the ecological, physiological, developmental, and genetic bases for population divergence can now be fully integrated for model and non-model organisms alike. From this body of work, case studies are accumulating to soon allow broad patterns to be identified concerning the relative importance of different mechanisms for the genesis of biodiversity. Although the mystery surrounding speciation may be gone, the thrill is not. Speciation is as important and as fascinating a question now, as ever, for understanding life.

### FRANCISCO AYALA

#### Darwin's Revolution

Darwin is deservedly given credit for the theory of evolution. He accumulated evidence demonstrating that organisms evolve over eons of time and diversify as they adapt to environments that are enormously diverse. Most important, however, is that he discovered natural selection, the process that accounts for the evolution of organisms and for their adaptive features; that is, their "design." The design of organisms is not intelligent, as it would be expected from an engineer, but imperfect and worse: defects, dysfunctions, oddities, waste, and cruelty pervade the living world.

Darwin's theory of evolution accounts for the design and diversity of organisms as the result of the gradual accumulation of spontaneous mutations sorted out by natural selection. Mutation and selection have jointly driven the marvelous process that, starting from microscopic organisms, has yielded orchids, birds, and humans. The theory of evolution conveys chance and necessity, randomness and determinism, jointly enmeshed in the stuff of life Darwin's fundamental discovery is that there is a natural process that is creative although not conscious.

### JEAN GAYON

#### History of Evolutionary Theories

Since 1859, evolutionary biologists have been haunted by the question of whether their conceptions are or are not

'Darwinian'. Although these terms are ambiguous, the repeated reference to Darwin has a theoretical signification.

Darwin settled a conceptual framework that has canalized evolutionary research over one-and-a half centuries. The structure of the two major aspects of this framework (the hypotheses of descent with modification and natural selection) are reconstituted and confronted with further evolutionary research, with special regard to the last 50 years.

In both cases, the article proposes a classification of the criticisms addressed to Darwin's two fundamental hypotheses, and compares the particular models of "descent with modification" and "natural selection" that Darwin defended with more recent models.

In *The Origin of Species,* the postulates underlying Darwin's hypothesis of "descent with modification" are expressed in a branching diagram, which has generated over time three major criticisms both among Darwinians and non-Darwinians: rejection of gradual modification, rejection of a conception of change exclusively concentrated at the level of the species, and more radical objections regarding the very idea that genealogy can be represented through a unique "tree". There may be several trees for several levels of integration of biological entities; lateral gene transfer and symbiosis also impose a network mode of representation.

As for the natural selection hypothesis, it has been criticized at two levels: the level that Darwin called the "mere hypothesis", and that of a "principle" able to explain and unify the whole theory of the history of life.

At the first level, three controversies have dominated since the 1960s: controversies over the neutral theory of molecular evolution, controversies over group selection, and controversies over the limits imposed by complexity and self-organization.

At the second level (Darwin's "well-grounded theory"), contemporary evolutionary biologists have challenged Darwin's idea that natural selection does account for as many "independent classes of facts" as adaptation, extinction, divergence, geological distribution of fossils, geographical distribution of species, relations between embryology and evolution, and patterns of classification.

Contemporary evolutionary biology admits that natural selection is the only acceptable explanation for adaptation, has raised serious doubts about the ability of natural selection to be an all-sufficient principle for the explanation of some or all the other classes of facts that Darwin explained through this principle.

In conclusion, the article examines the successive and ambivalent attitudes of Stephen Jay Gould towards "Darwinism". In 1980, Gould claimed that Darwinism was "dead". In his "Structure of Evolutionary Theory" (2002), he had a more nuanced appreciation, where Darwinism had not been either "extended" or "replaced", but "expanded". In Gould's terms, "expansion" means a reformulation of the fundamental principles of Darwinism, through generalization of the main Darwinian processes and addition of new principles.

The article concludes that the Darwinian framework has persisted, not under the form of the particular models of descent of modification and natural selection [Darwin] had in mind, but in the sense of high level heuristic postulates that have constrained and canalized the possible theoretical choices accessible to evolutionary biologists and paleontologists.

# Chapter 16

# Scott Gilbert: Evolutionary Mechanisms & Knish

*February 18, 2009*
*4:50 pm NZ*

> *"[T]he developing organism is a remarkable phenomenon. It respires before it has lungs, digests before it has a mouth, and creates itself anew from ordinary matter... Whereas the finished organism merely maintains its form, the embryo creates it."*
> – **Scott Gilbert**

You've got to love the chutzpah of a scientist like Swarthmore biologist Scott F. Gilbert, who (along with his students) once wrote a feminist critique of fertilization narrative, as well as his *joie de vivre* "moonlighting" as a piano player in a Jewish "mariachi" band called Knish. Gilbert can deliver a lesson in evolutionary biology with the verve of Lenny Bernstein. It's no wonder his textbook *Developmental Biology* is in its 8th edition and printed in a dozen languages.

He is also on the nine-man Scientific Committee for the upcoming evolution conference in Rome hosted by the Pontifical Council for Culture, March 3-7, where he'll present a paper titled:

**Evolutionary Developmental Biology: Evolution by Epigenesis** (Abstract follows)

In 1893, Thomas Huxley, wrote, "Evolution is not a speculation but a fact; and it takes place by epigenesis." Note that evolution's chief defender did not complete his sentence with the phrase "natural selection," for Huxley was interested in the generation of the diversity needed for natural selection.

That phase of evolution was regulated by development. Recent work has established five main mechanisms for the generation of anatomical diversity through changes in development, and this talk will review them and provide examples from the recent literature.

These mechanisms are:

(1) **Heterochrony** (changing the time or duration of developmental phenomena or gene expression)

(2) **Heterotopy** (changing the placement of developmental phenomena or the cell types in which a gene is expressed)

(3) **Heterometry** (changing the amount of gene expression in a manner sufficient to alter the phenotype)

(4) **Heterotypy** (changing the sequence of the gene being expressed during development)

(5) **Heterocyberny** (change in the "governance" of a trait from being environmentally induced to being genetically fixed)

These mechanisms have profound significance for how new traits can be generated, how they become integrated into developing organisms, and how they can become propagated through a population. It is argued that adding these developmental data and contexts provides a new and more complete theory of evolution, including a theory of body construction along with a theory of change.

Scott Gilbert is Howard A. Schneiderman Professor of Biology at Swarthmore College in Pennsylvania where he teaches embryology, developmental genetics as well as the history of biology.

Gilbert has a BA in biology and religion from Wesleyan, an MA in the history of science from Johns Hopkins (under Donna Haraway) and PhD in pediatric genetics from Johns Hopkins (under Barbara Migeon). He is married to Anne Raunio, an obstetrician and gynecologist, with whom he edited the book *Embryology: Constructing the Organism.*

He is a fellow of the AAAS and the St. Petersburg Society of Naturalists and has served as chair of the Division of Developmental and Cell Biology of the Society for the Integration of Cell Biology.

Some of Gilbert's awards include the Kowalevsky Prize in Evolutionary Developmental Biology, the Medal of Francois I from the College de France, a Guggenheim Foundation grant and an honorary doctorate from the University of Helsinki, among others.

He is currently investigating how the turtle forms its shell (ribs migrate to the dermis) with a grant from the National Science Foundation.

My recent phone conversation with Scott Gilbert follows.

**Suzan Mazur:** As you've said in your abstract for the upcoming Rome conference, evolution's chief defender Thomas Huxley noted that evolution takes place by epigenesis, and that diversity came before any natural selection could have. You present these main mechanisms for "generation of anatomical diversity": **heterochrony**, **heterotopy**, **heterometry**, **heterotypy** and **heterocyberny**. Why has the scientific establishment and the mainstream media not really embraced these ideas?

**Scott Gilbert:** It's hard to say whether the media has or not. The media's sources have to come from somewhere. That would be from evolutionary biology, from the scientists. Only recently has there been a cohort of evolutionary developmental biologists to answer reporters' queries.

I think that scientific paradigms are slow to change. Many evolutionary biologists, especially those who were trained in the 1970s and 1980s, do not feel there is a need for a theory of body construction to go along with the theory of change they have – which they think is a perfectly good one.

Evolutionary biologists have been very comfortable with the paradigm of the Modern Synthesis. The MS solves a lot of problems. It has modeled biology very well for the people who have asked it to model for them.

Given the limitations of the MS – for which you need interbreeding species – individuals with variations who can breed together, the program has worked remarkably well. However, with the MS the only way you could go to higher levels (above the species) was extrapolation.

What evolutionary biologists found was that evolution was predicated on mutation, recombination, and drift. It worked for what they were asking. Moreover, it made the extrapolation to higher levels possible.

Again, the notion that developmental evolutionary biology has is that you need a theory of body construction to supplement the Modern Synthesis, and then you could look at a theory of how to change body construction over time. You have to add developmental genetics to population genetics. This has been seen as unnecessary by many evolutionary biologists.

People like Ernst Mayr said this very explicitly. I quote him in the book I co-authored with David Epel, *Ecological Developmental Biology*. The publisher allowed me three historical appendices. One appendix is about how developmental biology was written out or ignored by evolutionary biologists.

**Suzan Mazur:** And then it became politically incorrect to question.

**Scott Gilbert:** Within its area it works. As long as evolutionary biology was defining itself this way, it had a beautiful model. The notion of natural selection in a breeding population works wonderfully. You add sexual selection to this. You add kin selection. You can take it in various directions. As long as there was a relatively parochial view of what evolution was and how it could be studied, the natural selection model and variation within a population worked really well. It was mathematical. It was based on mutations, which could be analyzed.

**Suzan Mazur:** But was it right?

**Scott Gilbert:** Was it right? For what it did, it was excellent . However, I'm on record in a 1996 paper saying that if the population genetics model of evolutionary biology isn't revised by developmental genetics, it will be as relevant to biology as Newtonian physics is to current physics.

**Suzan Mazur:** Do you still hold to that statement?

**Scott Gilbert:** Yes.

**Suzan Mazur:** So has money been wasted in the research that's been done?

**Scott Gilbert:** No. Not at all. I don't think it's been wasted.

**Suzan Mazur:** It could have been better spent.

**Scott Gilbert:** I think the priorities should now be changing, because until the 1990s we didn't have a theory of body construction.

That's the other thing. Evolutionary biologists in a way had every good reason to say we don't need a theory of body construction, or a theory of change in body construction, because the embryologists didn't have one.

One doesn't need a theory of body construction to talk about evolutionary change; but it becomes much richer when one has one. The thing that made it possible was DNA sequencing. This allowed us to compare genes between species, not merely within them. It showed that the genes involved in evolution were genes that are involved in constructing the body in the embryo.

**Suzan Mazur:** But why has the discourse been so sort of frat house - animal house? Why has it come down to that kind of conversation when these ideas seem to all work together?

**Scott Gilbert:** I don't know if I agree with you on frat house and animal house.

**Suzan Mazur:** Have you ever read PZ Myers?

**Scott Gilbert:** Okay. Yes. You'll get his view and you'll get Dawkins' view. And you'll get these people who have various agendas, and as you know all too well – science is done by people. No way of avoiding it. Thank goodness.

One has to show the data – and I think right now evolutionary developmental biologists do have a theory of the change in body construction. But before that you had to get certain things in place. You had to get proof of modularity in place, proof that changes in the enhancer regions can cause real selectable effects on morphology or on function. These things didn't come easy. They came about in piecemeal fashion. There wasn't coordination between laboratories.

The Society for Integrative and Comparative Biology formed the first evo-devo group in the year 2000. The first society for evolutionary-development in biology was established, I think, in 2006.

**Suzan Mazur:** Fairly recently.

**Scott Gilbert:** This is recent stuff. The first journals on evo-devo were also formed in the year 2000. So this is new. The coalescing of the field – sociologically speaking – is new. Yes, you can trace it back to the 1970s and how it's come along, but it really didn't gel until around 2000.

**Suzan Mazur:** But I find it peculiar that the two key evolution conferences have been organized outside the United States. One in Austria this last summer and the one in Rome in a couple of weeks. The day-long talks on evoutionary mechanisms at the Rome conference is extraordinary. Why is it the discussion taking place outside the United States? What statement is being made?

**Scott Gilbert:** What statement is being made? There have been conferences in the United States, but they've been generally smaller. There was a conference in 2001 at the SICB (Society for Integrative and Comparative Biology) meeting. It was the founding symposium for the evo-devo group at the SICB meeting.

There was actually one in 1991 at the American Society for Zoologists (the precursor of SICB). There were these smaller meetings. I think the one in 2001, which had international people in it, was in a way formative. The SICB meeting of 2001 brought American and European researchers together to talk. It also had a major historical component. Because, as you know, when you start a new field you have to go and find your forebearers and your heroes.

**Suzan Mazur:** Why didn't the media pick up on this in a major way as a breakthrough in science?

**Scott Gilbert:** It's a good question. The mainstream media does not attend the SICB meeting. The reporters from the scientific journals did write about it.The mainstream media to a large degree has a view of evolution that is Spencerian. One doesn't have to go far in the media to see that. It was David Brooks, I guess, in a recent *New York Times* piece who said that evolutionary science has gotten rid of any idea of or any vestige of generosity or communality. It's each for his own.

I think that the media has a competitive view of evolution, a very Spencerian and Hobbesian view. And they assume it's the whole notion of Darwinism.

**Suzan Mazur:** Red in tooth and claw?

**Scott Gilbert:** Exactly. I think that when you have these meetings where scientists are talking about mechanisms of getting variation and evolutionary-developmental biology, even more so ecological-developmental evolutionary biology, which talks about symbiosis and all sorts of group selection mechanisms – this doesn't fit into the media's notion of what evolution is.

**Suzan Mazur:** What machine do you think is at play preventing the media from reporting this? Is it economy-related?

**Scott Gilbert:** I think most of the people in the media have never had a course in evolutionary biology. They are perpetuating the view of evolution that they themselves are receiving from society (not from scientists). If they were to take an exam on evolutionary biology, I think that they would not pass. Even those people who say they believe in evolution probably wouldn't pass I don't want to ask them why do you believe in evolution? What's the data for it? Because they probably don't know. If the scientists say it's right, it's right.

**Suzan Mazur:** What do you think is the significance of evolutionary mechanisms being central to the Rome evolution conference?

**Scott Gilbert:** I think that having the notion of evolutionary mechanisms gets around having to be tied down to one paradigm. They're going fairly broadly. There are some things that are not being discussed. And other things that are. I think that in discussing the mechanisms, though, especially with the people who they have discussing the mechanisms, they are really casting a very wide net.

So that when you have people like Lynn Margulis talking about symbiosis and the origin of species, one hears a model of cooperation as opposed to a model of strict competition. When you have Stuart Kauffman and Stuart Newman talking about the physical antecedents and nongenetic evolution, you're dealing with things that are not in the normal paradigm of competitive natural selection.

Now I'm sure that natural selection will be here. You have people like Doug Futuyma talking, who will be discussing natural selection. But in a way natural selection has been proven so often and so definitively, I don't think it should be an issue. It's like discussing – are skeletons made of bones?

**Suzan Mazur:** So you don't agree that natural selection is more of a political term than anything else?

**Scott Gilbert:** I think natural selection occurs. And I think natural selection occurs within species. I don't think natural selection alone can explain how butterflies got their wings or how the turtle got its shell. (For that, you need to know developmental biology, as well.) But I think that once you have variation within species, then natural selection can work

**Suzan Mazur:** Do you think that natural selection should be relegated to a less important role in this whole discussion of evolution?

**Scott Gilbert:** Yes. I do. Natural selection has been touted by some scientists as the cause of variation. Here, however, I think development plays the major role.

But I'm not going to deny natural selection anymore than I would deny that friction is important in looking at the motion of objects in space – in our space anyway. Yes. It's there. There's natural selection. It works. And it works well within species.

To say that the fang of a rattlesnake evolved because of natural selection is absolutely correct, but also absolutely not sufficient. Because one has to say from what did the rattlesnake gets its poison fang? What modification of the salivary gland allowed it to be a poison gland? How did the change in development occur?

I had this argument with Michael Ruse. Michael Ruse and I had a wonderful exchange in *Biological Theory*. The fight that we had was over who were the heirs of Darwin. I think that's what it comes down to. Ruse is basically saying the Neo-Darwinists are the sole heirs of Darwin.

**Suzan Mazur:** And you're saying.

**Scott Gilbert:** I'm saying Darwin was an incredible guy who had all sorts of interests and that the evolutionary developmental biologists are actually asking a lot of the questions involving new species and how you get an adaptation that Darwin wrote about in his later books. *On The Origin of Species* was just his earliest book on this question, and then he dealt with variation and so forth in later books. In those books he realizes that natural selection is not enough.

**Suzan Mazur:** Well I think it's disappointing that the media has been focused on celebrating Darwin's birthday without a discussion of evolutionary mechanisms. Great to see that it's being talked about in Rome.

**Scott Gilbert:** I share your disappointment. When I sent my dues in to the National Center for Science Education, and I received a book called *Conceptual Issues In Evolutionary Biology*, there was nothing about evo-devo in it. And I said, "Well, these aren't the conceptual issues that grab me."

I was also disappointed in the *National Geographic* article. *National Geographic* would have been the perfect place for a current piece on new ideas in evolution. They ran a two-page feature on the evolution of views on evolution in biology, starting with Darwin. And the last thing – the last thing they show is a set of embryos. The caption to the side of it says current interest in the changes of genes bringing about changes in body plans is now being studied. The embryo story should have been the focus.

In the book that David Epel and I just published, we quote David Quammen: "Nature interests me because it's beautiful, complex and robust. Evolutionary theory interests me because it explains why nature is beautiful, complex and robust." Evolutionary mechanisms are really wonderful to study, and they tell us a great deal about the origin and maintenance of our biodiversity.

**Suzan Mazur:** Editors of news organization and their owners really underestimate the intelligence of the audience. They think they have to be general about information because that's all the public can grasp People want detail. People want to know what's happening. It's a miscalculation. More newspapers would be sold if the story were really reported.

**Scott Gilbert:** They like the conflict theory. I found the Brooks' article. It's the February 18, 2007 David Brooks *NYT* column – and I'm quoting: "From the content of our genes and the lessons of evolutionary biology it has become apparent that nature is filled with competition and conflicts of interest."

**Suzan Mazur:** Well he's a vehicle of the economic status quo.

**Scott Gilbert:** Of the right. Yes. But I think that's how evolution is taught. It comes around to what Huxley was saying about human nature, that we will use evolutionary biology to justify ourselves. And that in saying that nature is inherently amoral and self-interested – well, we're just part of nature. We justify our doing evil things because we say our genes made us do it. Darwinian selection. We've been selected to be competitive bastards. We don't usually hear about any other model, say, that we are the current pinnacle of the evolution towards cooperation.

**Suzan Mazur:** I think that some of this also has to do with so much of science being male-dominated. It's interesting that Lynn Margulis takes this perspective of cooperation. There was also an astrobiologist, a woman, at the World Science Festival this past summer who talked about "flow" – and questioned why we always have to think in terms of modules.

**Scott Gilbert:** Have you read Donna Haraway's book on this? Donna Haraway wrote a book called *Primate Visions*. It's about the gendered stories of primateology and how women have changed them. It's a beautiful book on how data is obtained and interpreted in a particularly important part of evolutionary biology, namely the ape-human interface. However, I wouldn't say that females (or feminists) are necessarily soft nurturers and that males must be hard-core militarists. It doesn't break down that easily.

I used to teach courses in feminist critique of science, and my students and I wrote a paper on feminist critiques of fertilization narrative, which was published back in the mid 1980s. I use feminist critique to show how some of our models of biology can be based on previously held models of human society.

In the fertilization narratives, the egg and sperm often become surrogates for women and men. (Remember the sperm sequence in *Look Who's Talking*?) The race of sperm and the female reproductive tract as passive conduit. Nothing can be further from the truth. No, that's a version of the old Hero Myth, where we are the descendents of the victor.

In actual fact, the female reproductive tract matures the sperm, and the first sperm getting to the egg do not fertilize it. Moreover, the egg activates the sperm before the sperm activates the egg. The egg is not the passive prize. But the public is told (repeatedly!) the story of conflict and prize winning

Similarly, our models of nature often come from pre-existing models of society. Russian biologists (even before the Communist Revolution), for instance, criticized early views of natural selection as being too rooted in British mercantile capitalism. One needs to step outside the cultural narratives to see if we are basing our views of nature on social norms. That's one of the reasons why international and humanities-based perspectives are important. They help stop science from becoming parochial.

But the important thing about scientific stories is that they are constrained by data. And that's an important aspect of the Rome conference. The Intelligent Design people are not being invited, and I suspect that this is because their stories are not constrained by data. They can tell any story they like (and they do.) Even after being shown that their statements have been disproved many times, the ID people still continue to tell these false stories, because the public can't tell if they are true or not. Scientists, however, have to limit their stories to that which the data allow.

# Chapter 17

# Evolution Sea Change? David H. Koch Weighs In

*February 24, 2009*
*12:27 pm NZ*

This story first appeared in *Archaeology* magazine. http://www.archaeology.org/online/interviews/koch/

It was an exquisitely warm, sunny February day and New York's groundhog had just bit the mayor, grabbing the headlines too. I made my way to the East Side, cutting through Barneys to the Madison Avenue offices of Koch Industries, Inc., the Kansas-based oil company. I had an appointment to talk about evolution with David H. Koch, a humanitarian with one of the world's great fortunes.

Not many people I've ever met have been to Tanzania's Olduvai Gorge – a place I had the thrill of visiting in 1980 – where Mary Leakey found *Zinjanthropus* (later renamed *Australopithecus*), and along with her team, the Hominin footprints at nearby Laetoli. So I was particularly delighted when David Koch opened our conversation by telling me of his expedition there in 1986 and shared some of his favorite things, such as a swatch of fossilized raindrops from Laetoli, which he held in his hands as if those drops were Faberge. Of all the possessions Koch might consider precious, who would have thought they'd be fossilized raindrops? But David Koch is committed to the investigation of human origins. And his philanthropy is serious.

Next year, the David H. Koch Hall of Human Origins opens at the Smithsonian National Museum of Natural History, where evidence of 6 million years of human evolution will be part of an interactive display that includes the Laetoli footprints and a reconstruction of Lucy. Visitors will be able to pass through a time tunnel to view early humans "floating in and out of focus," touch models of ancient human fossils as well as watch their own faces morph into those of extinct species. The Smithsonian display follows the creation of the American Museum of Natural History's David H. Koch Dinosaur Wing.

Richard Potts, director of the Smithsonian's Human Origins Program, explained about the new exhibition, "David's commitment to science and the study of human evolution will enable the Smithsonian to bring the latest discoveries in this field to the broadest audiences. The exhibition, still in the planning stages, encourages the public to explore the lengthy process of change in human characteristics over time. It also presents one of the new research themes in this field – the dramatic changes in environment that set the stage for human evolution. Although the subject can be controversial, the unearthed discoveries that bear on the question of human origins are a source of deep interest and significance for everyone to contemplate."

David Koch is Executive Vice President of $110 billion Koch Industries (he owns 42%) and CEO of its subsidiary, Koch Chemical Technology Group. He is often described as Manhattan's wealthiest resident, and contributes to Lincoln Center, Sloan Kettering Cancer Center, and the fertility clinic at New York-Presbyterian Hospital, to name a few. He is also is the principal private funder of PBS's *Nova* series.

Koch's BS and MS degrees are from MIT in chemical engineering. At 6'5" he also found some perspective away from the lab – shooting hoops. His MIT basketball plaque is displayed on his office trophy wall along with other treasures, including a framed replica of Lucy's hand.

I asked him about Olduvai, human origins, changes in evolutionary thinking, and more.

**David Koch:** It [Olduvai] is unbelievable. As far as you can see there are animal bones like this everywhere! When you were there I'm surprised they didn't show that to you.

**Suzan Mazur:** There were regional tensions at the time I flew into the Gorge from Nairobi. It was 1980. In fact, the border was officially shut down between Kenya and Tanzania. Authorities in Dar es Salaam gave me permission to land for a few hours, and only to interview Mary Leakey for *Omni* magazine. The pilot of a single engine Cessna flew me in. We couldn't find the Gorge. It was the dry season and our maps were from the wet season. Had to circle three times before locating it. I was getting sick. Then we found an opening in the terrain, Olduvai, and dove in. Mary Leakey drove out to meet us. Introduced us to her four dalmatians. Made us some lunch – macaroni and cheese casserole, and we talked.

**David Koch:** My friend Don Johanson organized our expedition in 1986.

**Suzan Mazur:** He did a two-part documentary for PBS.

**David Koch:** Three-part. It was on human evolution. Don was the host of it.

**Suzan Mazur:** You also supported his institute.

**David Koch:** I still am supporting it, I'm on the board there.

**Suzan Mazur:** He found Lucy.

**David Koch:** Yes. When I got there they had discovered a Hominin's bones. They left them in the earth, waiting for me to arrive. And then when I arrived, they let me pull them out of the ground, which was kind of fun.

**Suzan Mazur:** Well there's a conference coming up at the Vatican in March.

**David Koch:** On why creationism is real?

**Suzan Mazur:** The premise is that "issues surrounding evolutionary biology merit a careful and serious reconsideration."

**David Koch:** Oh, so they're opposed to it.

**Suzan Mazur:** No, they're moving deeper into a discussion of evolutionary science. They're going beyond. . .

**David Koch:** I've always felt devout religious advocates believe human evolution and evolution in general are incompatible with the concept of a divine God.

**Suzan Mazur:** The Vatican is saying the two can co-exist and that religion should in no way be a scientific theory and evolutionary science should not be dogma. The interesting thing is that the Vatican has invited experts on these other evolutionary mechanisms aside from natural selection. People like Stuart Newman (A "pattern language" for evolution and development of animal form), who I've interviewed in *Archaeology* magazine. Lynn Margulis (symbiosis), who was awarded the President's National Medal of Science. Stuart Kauffman (evolution and complexity) – a big name. Colin Renfrew's going to be there. It's a huge gathering of people. Francisco Ayala.

**David Koch:** Do they have an equal number of creationists presenting?

**Suzan Mazur:** At the end of the program they've got philosophers and one or two people talking about intelligent design but no creationist or intelligent design people presenting papers.

**David Koch:** That's interesting. It's hard to believe the Catholic professionals would support the evolutionary theory of Charles Darwin.

**Suzan Mazur:** There's a big shakeup going on, which is what I've been reporting.

**David Koch:** After all Galileo was imprisoned for years for saying the world was round. Evolution's a hell of a lot more extreme than Galileo's concept.

**Suzan Mazur:** There's been a huge debate this past year particularly. I'm not referring to evolution vs. creation. What I've been covering involves other mechanisms of evolutionary change aside from Charles Darwin's natural selection. Some of the most savvy scientists would like to see natural selection relegated to a lesser role.

I've written an expose of the evolution industry.

**David Koch:** Are you an evolutionist or a creationist?

**Suzan Mazur:** I'm an evolutionist. I've been talking to scientists who are going deeper into the investigation of evolutionary science. Biology is looking to physics now for answers about evolution. They've discovered as many genes as they're going to find for humans – 20,000-25,000.

**David Koch:** Can I interject a little story? I'm on the board of MIT and one of the main contributors at least in the field of biology and cancer research.

About a year and a half ago I went to a seminar where the speakers were some of MIT's most brilliant and highly acclaimed people. They were talking about the latest and greatest research that's going on there. One speaker after another – these are outstanding, world class scientists, Nobel Prize winners in some cases. On the same faculty they differed enormously in the number of genes that have been discovered. There's no consensus.

**Suzan Mazur:** There's a range of 20,000 - 25,000.

**David Koch:** It went down to as little as 15,000 genes and some of them went up to 30,000 genes. Nobody really knows.

**Suzan Mazur:** And they don't even know what the gene is. That's the discussion now. But since you only have 35 minutes, can we begin with more formal questions?

**David Koch:** Sure.

**Suzan Mazur:** You have enormously influenced the public's understanding of science through your support for programs on PBS and *Nova*. You've given to the Institute for the Study of Human Origins and the Louis B. Leakey Foundation. You've funded a dinosaur wing at the American Museum of Natural History. Next year the David H. Koch Hall of Human Origins opens at the Smithsonian Museum. You founded a cancer center at MIT. You're on the board of the Cato Institute and you and George Soros helped to finance the ACLU's successful push to deal with the PATRIOT Act, among many other humanitarian gestures and generosities. You are a major donor to the arts.

What are some of your other community interests and concerns now?

**David Koch:** I give a great deal of money to sponsor research and facilities for research in the effort to find cures for various types of cancer. I, myself, suffer from prostate cancer which I found I had almost 17 years ago. So it's a great personal interest of mine.

**Suzan Mazur:** They've come a long way with treatment. Are you okay?

**David Koch:** I'm doing fine. I still have the cancer.

**Suzan Mazur:** You have to monitor.

**David Koch:** Yes. Over the years I've developed strong relationships with quite a number of outstanding cancer research institutes and centers. And during the time I've spent with those organizations and with the funds that I've provided, I've moved the field of cancer research substantially forward. I feel very proud of that.

**Suzan Mazur:** Are you involved in any way in the editorial content of *Nova* programs on evolutionary science?

**David Koch:** No I am not. I've been following the *Nova* series ever since it first came on the air. I'm a great admirer.

**Suzan Mazur:** But you stay out of the content.

**David Koch:** That's right. The quality of the work they do is outstanding. And I think it stands rigorous analysis. It's the latest and greatest. And they present it so beautifully that the average lay person can understand it quite easily.

**Suzan Mazur:** You ran for U.S. vice president on the Libertarian ticket in 1980, considered the most successful Libertarian presidential ticket ever, getting roughly a million votes. What role do you think politics should play in educating the public about evolution?

**David Koch:** That's an interesting question. I think politicians should really stay out of it and allow scientists to present the facts and discoveries. I hate to see it politicized.

It's like saying what role should politics play in, for instance, religion? I think it should be up to individuals to decide what they believe. So often politicians are totally uninformed about scientific facts.

**Suzan Mazur:** And what about the local school boards?

**David Koch:** There again, the school boards should not have rigorous control over that subject. I think science teachers should be allowed to teach it very openly, without restrictions on what they can say.

**Suzan Mazur:** As a man committed to the principles and practices of freedom, including scientific freedom, and as a scientist yourself with degrees from MIT in chemical engineering – is it your perspective that we are now witnessing a sea change in evolutionary thinking? That even as the global celebration begins for Charles Darwin's 200th birthday, the man who brought us the theory of evolution by natural selection 150 years ago – Darwinian selection, or survival of the fittest, is now being viewed by serious evolutionary scientists as not enough to explain our existence?

To quote from my interview several months ago with NASA astrobiologist Chris McKay, who was featured in the recent *Nova* Mars documentary you helped underwrite: "Something had to precede Darwinian natural selection. The Darwinian paradigm breaks down in two obvious ways. First, and most clear, Darwinian selection cannot be responsible for the origin of life. Second, there is some thought that Darwinian selection cannot fully explain the rise of complexity at the molecular level." So the question is: Is it your perspective that we are now witnessing a sea change in evolutionary thinking?

**David Koch:** No. I don't think it's a sea change. The sea change occurred back when Darwin published his evolutionary theories, backed up by massive, overwhelming evidence. What's happened since is that there's been a rather steady progressive acceptance of the concepts of evolution in the general public. It's amazing to me that in America a large faction of the population still doesn't believe in it.

**Suzan Mazur:** But the point is that Darwin started with life. He addressed what happens once you have life. He didn't address the origin of life. That's what Chris McKay, the NASA astrobiologist is saying.

**David Koch:** Scientific knowledge of early life was not something that had been discovered when Darwin was alive.

A huge amount of knowledge of how life might have begun has now been determined.

**Suzan Mazur:** Much of the media and scientific community appear to be stuck in the debate on evolution vs. creationism. A recent Gallup poll in America revealed that two-thirds of Republicans questioned rejected Darwin's theory and a majority of Democrats and political Independents accepted it. What is consistently ignored by pollsters and the media is the evolutionary mechanisms aside from Darwinian natural selection.

More sophisticated evolutionary thinkers are now saying natural selection is not the most important mechanism of evolutionary change. I'm talking about scientists who are funded by the National Science Foundation, not kooks.

*What Darwin Got Wrong* is a forthcoming book co-authored by Jerry Fodor, one of America's most celebrated philosophers, who argues that at the end of the story "it's not going to be the selectionist story". A Swedish cytogeneticist, Antonio Lima-de-Faria, who's been knighted by the king of Sweden for his scientific accomplishments, has noted that "there has never been a theory of evolution."

In fact, there is a parallel celebration this year of the 200th anniversary of Jean-Baptiste Lamarck, the scientist who was onto the idea of evolution before Darwin. New York Medical College cell biologist Stuart Newman has said publicly he believes that "over the next couple of decades Lamarck's way of looking at things [the inheritance of acquired characteristics] will be more incorporated into mainstream biology."

Would you comment?

**David Koch:** Well I'm not an authority on all those details. I have a general working knowledge of evolution. I'm not competent to challenge some of the claims of those folks.

**Suzan Mazur:** This is a big debate, which the media is not covering. It's reached a crescendo and a lot of people are saying there's a sea change happening. Some of the evolutionary mechanisms being discussed, which relegate natural selection to a less important role, include self-organization – where cells organize themselves into more complex structures. The concept of morphogenetic fields, a developmental grid guiding development, is something Mount Holyoke paleontologist Mark McMenamin and Stuart Pivar have been investigating, identifying the famous Seilacher Namibian fossil that was part of Steve Gould's *Scientific American* article as a flattened morphogenetic torus, a metazoan creature.

**David Koch:** I'm not sure what the significance of that discovery is. It seems to me what's amazing is how much Darwin got right 150 years ago. It's staggering what he got right. He got enormously more right on evolution than what he got wrong.

**Suzan Mazur:** These people aren't questioning the concept of evolution. What they're saying is that there needs to be more, that we need to go beyond Darwin for answers. There's also something called saltational mechanisms which produce abrupt evolutionary change, that is – jumps – where one form rapidly replaces another. Niche construction where organisms invent their habitats rather than being selected.

**David Koch:** There's been a fine-tuning of Darwin's evolutionary theory, there's no question.

**Suzan Mazur:** Then there's epigenesis, where a chemical layer is laid down on top of the genes resulting from various stresses on the organism, and the resulting traits (including disease) can be passed on without changes to the DNA. A kind of neo-Lamarckian concept.

What I'm asking is, should the media, and in particular, PBS, focus on these better ideas of how evolution occurred and by enlightening the public, help stop the fighting about "old science"?

**David Koch:** As more and more knowledge is developed over time as to how evolution at the molecular level is driven, how it works – I think it's a very important responsibility of programs like *Nova* to continually update the public on the latest findings. I certainly agree with that.

**Suzan Mazur:** That's good to hear.

**David Koch:** If there's a difference of opinion between one scientist and another, or a third scientist and that debate can help clarify what's going on in the field of evolution – I think it's important to publish that and discuss it on those kinds of programs.

**Suzan Mazur:** As I mentioned earlier, next month in Rome the Vatican (Pontifical Gregorian University in collaboration with Notre Dame) will host an international conference open to the public called: "Biological Evolution Fact and Theories: A Critical Appraisal 150 Years After *The Origin of Species*".

One whole day out of three days will be devoted to a discussion of these evolutionary mechanisms with scientists, some of whom I've already noted, Stuart Kauffman, Lynn Margulis, Robert Ulanowicz, Scott Gilbert and others presenting papers. This comes on the heels of the Altenberg 16 scientists meeting last July outside Vienna to kick off what they now call the Extended Synthesis which updates the neo-Darwinian or Modern Synthesis which was last updated 70 years ago.

So far we have not seen these kinds of groundbreaking meetings taking place in America. Speakers at the annual AAAS meetings are organized by Eugenie Scott from the National Center for Science Education, who told me at the

Rockefeller Evolution conference in May that her organization does not recommend textbooks for schools if those texts include a discussion of self-organization because it is confused with intelligent design. In effect, NCSE is recommending old middlebrow science for kids. There's a "cycle of submission" at play here.

[NCSE has responded to this article saying that Eugenie Scott has organized symposia, not speakers, at AAAS meetings, and "NCSE does not recommend specific textbooks at all, although we encourage textbook publishers to ensure that their treatment of evolution is extensive, pervasive, and up-to-date, and we oppose the use of textbooks that treat creationism as scientifically credible."]

Do you have any interest in supporting an evolution conference in America along the lines of what the Vatican or the Austrians have done? Also, do you have any interest in creating a foundation specifically for the investigation of these other mechanisms of evolution?

**David Koch:** It's like debating how many angels can dance on the head of a pin. I don't think there's much practical relevance to all this. Life started somehow. The details of how it started I don't think anyone will ever be able to prove.

I think my talents and fortune could be enormously better spent on developing cures for diseases like cancer. For me to worry about these highly theoretical arguments rather than try to cure these horrible diseases? Cancer kills half a million people a year. That's a far better use for my money than this kind of academic theoretical debate.

**Suzan Mazur:** Stuart Newman was the focus of considerable media attention 10 years ago for his attempt to patent a part human-part animal chimera.

**David Koch:** A hybrid.

**Suzan Mazur:** He tried to patent it to show the dangers of the commercialization and industrialization of organisms.

What he's saying now is that because the public does not have an up-to-date understanding of how evolution happens – partly because science is stuck in the Darwinian model – people are less likely to object to genetic engineering experiments because the thinking is that there will just be minor changes. But according to Newman, there's potentially a huge danger because jumps can happen. The evolution may not be so gradual. Genetic engineering seen in that light looks considerably riskier.

**David Koch:** There's some infinitesimal probability that could happen. But it's hardly worth worrying about it. I'm more worried some strange disease could show up on our shores from Africa like AIDS did and could kill millions of people. The epidemic of another strange disease. That to me is enormously more likely to happen than some of these wild, far out concerns of evolutionary study.

**Suzan Mazur:** One of the problems is that there's been a big emphasis on genetics at the expense of the physical sciences, even though scientists still don't understand what the origin of the gene is. In fact, scientists don't even agree on what evoluton is.

Big money has been thrown at genetic research since the 1930s, first by the Atomic Energy Commission, over concern for mutation caused by exposure to radiation. Now that we've found all the human genes we're going to find, there's been a kind of U-turn back to embryology to see what else is happening. A shift from the gene-centered perspective of evolution to non-centrality of the gene. A directional shift in biology back to physics and chemistry.

It's a deeper approach to understanding evolution. They're not kooky ideas. The concept of self-assembly, for instance, where you put certain chemicals into a beaker or test tube, shake it up and vesicles form.

**David Koch:** Natural connections you're saying. Well yes, that's how the human egg grows into an adult.

My wife and I are a major supporter of a fertility clinic in New York and it's incredible what they've done to create normal adults from infertile people. They have an understanding of how eggs develop, that's why they've been so successful.

The head guy over at New York Presbyterian Hospital is responsible for about 15,000 normal healthy babies. I used to think ibn Saud was a hell of a guy. He was the founder of modern Saudi Arabia and he had 700 children. But I told the guy at NYPH, you're up to 15,000 and counting. You've got ibn Saud beat by a mile.

## Chapter 18

# Lynn Margulis: Intimacy of Strangers and Natural Setection

*March 16, 2009*
*5:59 pm NZ*

> *"Lynn Margulis is an example of somebody who didn't follow the rules and pissed a lot of people off. She had a way of looking at symbiosis which didn't fit into the popular theories and structure. In the minds of many people, she went around the powers that be and took her theories directly to the public, which annoyed them all. It particularly annoyed them because she turned out to be right."*
> – **W. Daniel Hillis, *The Third Culture***

While Eastman Professor Lynn Margulis clearly doesn't have time on her hands at Oxford University's Balliol College where she's spending the year away from her other job as Distinguished University Professor of Geosciences at the University of Massachusetts-Amherst – I did actually run out of tape talking with her in round one of our conversation, barely scratching the surface on symbiosis ("new species evolve primarily through the long-lasting intimacy of strangers"), the evolutionary concept that brought her the Presidential Medal of Science in 1999. Margulis says that as far as "survival of the fittest" goes, it's a "capitalistic, competitive, cost-benefit interpretation of Darwin" and that even banks and sports teams have to cooperate to compete. She sees natural selection as "neither the source of heritable novelty nor the entire evolutionary process" and has pronounced neo-Darwinism "dead," since there's no adequate evidence in the literature that random mutations result in new species.

Margulis takes a holistic view of evolutionary science, and her U. Mass. lab page notes that their work "seamlessly" involves microbiology, cell biology, genetics, ecology, "soft rock" geology, astronomy, astrobiology, atmospheric sciences, metabolic organic and biochemistry.

This year on Darwin's 200th birthday anniversary (Feb. 12), she was awarded the Darwin Wallace medal by the Linnean Society of London. She was elected to the National Academy of Sciences in 1983, served as chair of the NAS Space Science Board Committee on Planetary Biology and Chemical Evolution, was elected to the Russian Academy of Natural Sciences, the World Academy of Art and Science and the American Academy of Arts and Sciences, among other honors.

Lynn Margulis told me that when she wrote her book *Symbiosis in Cell Evolution: Microbial Communities in the Archean and Proterozoic Eons* she was entirely ignorant of the Russian work of Boris Mikhailovich Kozo-Polyansky (1924) and his predecessor's concepts of symbiogenesis. She said she also knew little of the American antecedents (*e.g.*, Ivan E. Wallin's *Symbionticism and the Origin of Species*, 1927). Margulis said she simply had read with great interest Columbia Professor E.B. Wilson's tome "The Cell in Heredity and Development" (1925). Yet her conclusions closely resembled those of Kozo-Polyansky and other unknown symbiogenesis-championing predecessors albeit with modern genetic, biochemical and paleontological information.

She has co-authored seven books with Dorion Sagan, her son from her first marriage (at 19) to the late astronomer Carl Sagan. Those books include: *Symbiotic Planet: A new look at evolution; Acquiring Genomes: A theory of the origins of species; What is Sex?; What is Life?; Mystery Dance: On the evolution of human sexuality; Microcosmos: Four billion years of evolution from our microbial ancestors; Origins of Sex: Three billion years of genetic recombination.* With M.J. Chapman, her close colleague and former student, she's written *Kingdoms & Domains: An illustrated guide to the phyla of life on Earth*, now in its 4th edition.

She adores many of her colleagues, describing them as "marvelous!", "wonderful!" "superb!". And she advised that she would have "no time" to talk with me as soon as her daughter, son-in-law and three grandchildren arrive for a visit.

Lynn Margulis spent 22 years teaching at Boston University prior to her current faculty positions. She has a BA from the University of Chicago, an MS in zoology and genetics from the University of Wisconsin and a PhD in genetics from the University of California, Berkeley.

But she didn't start out this way. She was born on the south side of Chicago to a non-science family, had a wild streak and admits (most women won't) that she loved to chase and be chased by guys. She married two of them.

Our recent phone conversation follows a slightly improved abstract of the paper she presented in Rome last week at the "Biological Evolution Fact and Theories" conference organized jointly by the Jesuits, Pontifical Gregorian University (Rome) and the University of Notre Dame (Indiana).

## Origin of Evolutionary Novelty by Symbiogenesis

Whereas speciation by accumulation of "random DNA mutations" has never been adequately documented, a plethora of high-quality scientific studies has unequivocally shown symbiogenesis to be at the basis of the origin of species and more inclusive taxa.

Members of at least two prokaryotic domains (a sulfidogenic archaebacterium, a sulfide-oxidizing motile eubacterium) merged in the origin of the earliest nucleated organisms to evolve in the mid-Proterozoic Eon (c. 1200 million years ago).

Such a heterotrophic, phagocytotic motile protoctist was ancestral to all subsequent eukaryotes (*e.g.*, other protoctists, animals, fungi and plants).

The defining seme of eukaryosis, the membrane-bounded nucleus as a component of the karyomastigont, evolved as *Thermoplasma*-like archaebacteria and *Perfilievia-Spirochaeta*-like eubacteria symbiogenetically formed the amitochondriate LECA (the Last Eukaryotic Common Ancestor). Their co-descendants (that still thrive in organic-rich anoxic habitats) are amenable to study so that our videos of them will be shown here.

There are no missing links in our scenario. Contemporary photosynthetic (green) animals (*e.g., Elysia viridis, Convoluta roscoffensis*), nitrogen-fixing fungi (*Geosiphon pyriforme*), cellulose digesting animals (cows, *Mastotermes darwiniensis* termites) and plants (*Gunnera manicata*) make us virtually certain that Boris Mikhailovich Kozo-Polyansky's (1890-1957) analysis (*Symbiogenesis: A New Principle of Evolution*, 1924) was and still is correct.

Symbiogenesis accounts for the origin of hereditary variation that is maintained and perpetuated by Charles Darwin's natural selective limitations to reaching the omnipresent biotic potential characteristics of any species.

**Suzan Mazur:** What is the significance of the Rome evolution conference and why was it limited it to US-European papers?

**Lynn Margulis:** I didn't know it had been. You mean no Chinese or Japanese?

**Suzan Mazur:** There are no Russian, Chinese, African, Indian or Japanese presenters listed.

**Lynn Margulis:** This is not a policy of limitation, this fact resulted from historical circumstances.

It must be deeply understood that the term "evolution," which is not used by Charles Darwin – he called the process "descent with modification" – is Anglo-Saxon. It is very much a British-American "take" on the history of life, traditionally limited to Anglophones.

Most English-speaking scientists think in hushed hagiographic terms when they mention Charles Darwin, comparable to English thought about physics before Einstein when Newton was the only game in town. It's a very English nationalist phenomenon, especially as Darwin was later interpreted.

**Suzan Mazur:** Do you think the Rome conference organizers had that in mind when they were inviting papers?

**Lynn Margulis:** No I don't think so. It probably didn't even occur to them that the guest list on their "international meeting" might strike some as racist!

The Chinese and the Navajos lack any tradition in evolution, although they both enjoy superb medicine (healing) traditional practice.

Professor Tom Glick, a former colleague of mine at Boston University – he's wonderful – wrote a book, *The Comparative Reception of Darwin,* with chapters by country.

A joint student of ours suggested the study needed a chapter on the Chinese reception of Darwinism. The book has a chapter on Japan, Latin American coverage, Spain, many countries – on how Darwinism was perceived and received in the century between 1859 and about 1970.

This young man, a doctoral candidate in the history of science, went to China for a year and discovered no tradition of Darwinian evolution there. He ended up studying aspects of Chinese medicine.

Also, my colleague Tacheeni Scott, a fine cell biologist, a Navajo, told me that his culture has no concept whatsoever of evolution. They just have no tradition.

**Suzan Mazur:** But there is significant research on evolution taking place in India and Japan. I haven't looked at African evolution studies but I did interview scientists in Africa in the 1980s for *Omni* magazine – they were trained by the Soviets, so there must be important African thinking about evolution.

**Lynn Margulis:** Most of Africa was colonized by Europe. Let's put it this way. In the Russian equivalent of the *Encyclopedia Britannica,* some 250 pages describe symbiogenesis – in an evolutionary context of course. In Darwinian evolutionary books published in and before 1982 as part of the centenary activity of Darwin's death in 1882, there is zero on symbiogenesis.

You have a point. Certain countries are expected to be excluded because they lack traditional study of evolution.

**Suzan Mazur:** The theme of the Rome Conference is "Biological Evolution Facts and Theories: A Critical Appraisal 150 Years After *The Origin of Species*". Would you comment on how funding colors evolution fact and theory?

**Lynn Margulis:** I will give you a specific example. In perhaps about 1980, Harvard professor Richard Lewontin – you know him?

**Suzan Mazur:** Yes. I've interviewed him.

**Lynn Margulis:** And the late Margaret Dayhoff. . . She was a protein biochemist who started the use of protein sequence information to reconstruct evolutionary history. She co-authored a marvelous series of books on *Protein Sequence and Structure* in the early 1970s with Richard Eck. Their handbook collected all the evolutionary information at the time, which wasn't much. Dayhoff *et al.* realized that the kinds of data they were getting could only be comprehended in the light of evolution.

The late paleontologist and Harvard professor Elso Barghoorn was involved too. There were four or five of us who, by correspondence exclusively, realized that the major issue in all our research – whether it was electrophoresis or the microfossil record – was the evolution story. So we talked about ways of putting pressure on the National Science Foundation to set up an evolution section. This clearly was in NSF's (not NIH's) purview. Dayhoff was funded by a chemistry section of the National Institute of Health. Barghoorn was supported by a geology section of NSF and by NASA.

We wrote a carefully honed letter that said a very strong set of researchers existed who consider their primary activity "evolution" and yet their methods are very different. We proposed that our efforts be joined. This would lead to reduction of redundancy and save money for the funding agencies. It would probably further evolution science more than anything else to construct an evolution program. We investigators would not have to prevaricate about our interests. We said it politely.

I think I sent the letter in. I did not receive an answer from them for over a year, perhaps for two years. Then out of the blue, long after I figured there would never be an answer and had entirely forgotten, I received an answer from the NSF.

Remember Wisconsin Senator Gaylord Nelson who ridiculed the NSF saying the NSF funded work on the left toe of spiders or something? He was just trying to be sensible. I can understand.

Anyway, I deduced that NSF scientist-bureaucrats were conflicted about our letter. The woman assigned to answer us wrote to say there were so many American citizens opposed to evolution that if the NSF put chemistry, geology, *etc.* into a single evolution division, it would be like sticking out our heads to be chopped off. Such a proposal, no matter its intellectual validity, would surely not fly! She said the NSF thought it would strengthen evolution science by avoidance of the word "evolution" and not by centralizing research activities.

**Suzan Mazur:** I've been critical about the NAS publication *Science, Evolution, and Creationism.* It's a light treatment of the subject.

**Lynn Margulis:** Mealymouthed probably.

**Suzan Mazur:** Would you say there's an evolution sea change taking place?

**Lynn Margulis:** Tell me more please.

**Suzan Mazur:** For instance, at the Rome conference there's a full day of talks on "evolutionary mechanisms".

Stuart Kauffman, one of the scientists presenting a paper on evolutionary mechanisms, told me in an interview about a year ago "there are some physicists who are asking questions like: Is natural selection an expression of some more general process?" That "it's all up in the air".

Richard Lewontin told me that natural selection occurs.

Antonio Lima-de-Faria says natural selection's a political term not a scientific term.

Philosopher Massimo Pigliucci told me publicly that it's both politics and science.

**Lynn Margulis:** Who first said it was a political term?

**Suzan Mazur:** Antonio Lima-de-Faria, the cytogeneticist from the University of Lund.

**Lynn Margulis:** From Lund. He's Portuguese.

**Suzan Mazur:** Right.

**Lynn Margulis:** Good, good. Is he coming to Rome?

**Suzan Mazur:** He is not. I've had lots of dialogue with him.

Stuart Newman says natural selection should be relegated to a less important role in evolutionary science.

Stan Salthe says "Oh sure natural selection's been demonstrated . . . however . . . it has rarely if ever been demonstrated to have anything to do with evolution in the sense of long-term changes in populations. . . ."

**Lynn Margulis:** That's really what Salthe said?

**Suzan Mazur:** Yes. And he said that "the import of the Darwinian theory of evolution is just unexplainable caprice from top to bottom. What evolves is just what happened to happen."

Philosopher Jerry Fodor, who's co-writing a book *What Darwin Got Wrong* with physicist and linguist Massimo Piattelli-Palmarini, told me that "whatever the story turns out to be, it's not going to be the selectionist story."

Can you shed some light on what's going on regarding the status and meaning of natural selection?

**Lynn Margulis:** I think I see the problem clearly. There is absolutely no doubt that natural selection itself can be measured every minute of the day in every population of organisms. Darwin was brilliant to make "natural selection" a sort of godlike term, an expression that could replace "God", who did it – created life forms. However, what is "natural selection" really? It is the failure of biotic potential to be reached. And it's quantitative.

Biotic potential is the intrinsic ability of any population to overgrow its environment by production of too many offspring. Whether born, hatched, budded or sporulated, all organisms potentially produce more offspring than can survive to reproduce themselves. Natural selection is intrinsically an elimination process. I'll give you some specific examples.

My favorite one – I show this in a film and people just gasp. An ordinary bacterium – *Proteus vulgaris* – divides at the rate of every 15 minutes.

I have a time-lapse view of *Proteus vulgaris* where I show two hours of growth – 1, 2, 4, 8, 16, 32, 64, etc., until it fills the screen. I explain that if *Proteus vulgaris* continued to grow at this rate, not once a minute or once every 10 seconds, but once every 15 or 20 minutes, *i.e.*, the way it really grows when it's not limited, this bacterium would reach the mass of the Earth over a weekend.

It's easy to show that the biotic potential measured as "number of offspring per unit time" (convertible of course into its equivalent "number of offspring per generation") is never reached. Ever.

Darwin said the whole Earth could be covered by the progeny of a single pair of elephants.

Everybody knows that ancient chess story, one rice grain, two rice grains, four rice grains – you know what I'm talking about – the whole kingdom of rice at the end of the chess board = $2^{64th}$!

Apparently, a maximum of 11 dachshund puppies can be born per litter. They have 3 litters a year, and therefore their biotic potential is: 33 puppies a year.

Let's take a human example. For years the eldest woman reported to have delivered a live infant was a 59-year-old. The highest biotic potential described was 22 children per one single couple. This was the measurement of "human biotic potential": "22 children per couple per generation", and approximately equivalent to 22 children per 25 years.

Recently a Brazilian newspaper reported a couple, same Mom, same Pop, who had 32 children! Now each of those children, I can bet you, I don't have to bet you, did not live to produce 32 children in the next generation. That has never occurred in the history of mankind. That point is – if you return 25 years later to the same Brazilian village, the original two parents will have been replaced, not by 32 but by another two descendants. The others died, moved away or, most likely, most failed to reproduce. That is simply elimination by "natural selection", the failure of biotic potential to be reached. But that's all. Simply no population ever reaches its biotic potential for long enough to do anything but measure it!

What then is natural selection? Natural selection is the failure to reach the potential, the maximum number of offspring that, in principle, can be produced by members of the specific species in question. This has been shown zillions of times in zillions of organisms.

**Suzan Mazur:** So you don't agree that natural selection has rarely if ever been demonstrated to have anything to do with evolution in the sense of long-term changes in populations.

**Lynn Margulis:** Of course not. We have to unpack that misstatement. Growth is not simply enlargement by intake of food. It is no single process. In metazoans growth involves, at least intracellular motility including mitosis, protein synthesis, ATP energy generation, oxygen respiration, water intake and retention, salt balance, development and cell differentiation, and other related processes. The term "growth" tends to be slighted and misrepresented but it is far worse with the word "evolution". The evolutionary process, intrinsically multi-componented, tends to be misunderstood by most people; it is often not even properly presented by those who purport to teach it! I think I understand it and can unpack it with complete equanimity. Natural selection occurs all the time. But natural selection as an elimination process, as failure to reach biotic potential, is not the issue.

**Suzan Mazur:** Salthe's saying natural selection in terms of long-term changes in populations.

**Lynn Margulis:** I claim that long-term change also has been demonstrated. Such change over time is what the whole fossil record is about

**Suzan Mazur:** He said it's been demonstrated but rarely. . .

What about Stuart Kauffman? He said natural selection may be "an expression of a more general process."

**Lynn Margulis:** They are arguing about the entire evolutionary process. They are confused about its separable, measurable components. Darwin's claim of "descent with modification" as **caused by natural selection** is a linguistic fallacy. They talk as if there were one single cause. As if natural selection were **the** cause. Although stated in your quotes, they respond to that which is left undefined. They do not respond to evolutionary evidence, to the results of the evolutionary process as documented in the fossil record. They tend to be ignorant about sedimentation, stratigraphy, taphonomy, diagenesis and other natural processes relevant to interpretation of direct fossil evidence for life's evolution.

Darwin wrote about the Struggle for Life and attributed change to Natural Selection. He made it easy for his contemporaries to think and verbalize Mr. Big Omnipotent God in the Sky up there picking out those He wants to keep. He has been conceived of as The Natural Selector, He throws the others away.

**Suzan Mazur:** Your investigation of holistic science has revolutionized thinking about evolution. Your lab page carries the statement: "Our science seamlessly involves microbiology, cell biology, genetics, ecology, "soft rock" geology, astronomy, atmospheric sciences, metabolic organic and biochemistry."

You've been critical of evolutionary science being too focused on animal investigation and not looking enough to more than two billion years that preceded the origin of animal species. What are your thoughts about what may have preceded biological evolution? Do you find Antonio Lima-de-Faria's idea interesting about atomic, chemical, mineral, chemical levels of evolution?

**Lynn Margulis:** Evolution is not the appropriate word. "Evolution" just means change through time. But yes there has been "change though time" at many levels.

**Suzan Mazur:** Are you saying you agree there are four levels?

**Lynn Margulis:** I don't know what Lima-de-Faria's four levels are. I'm familiar only with his good work on cells from years ago.

**Suzan Mazur:** What about his ideas about biological periodicity, where he describes the flatworm, for instance, having male genitals as developed as those of the human male, the placenta turning up in assorted species across time, along with traits like luminescence, the capacity to fly, etc.

**Lynn Margulis:** I assume he's talking about convergent evolution.

**Suzan Mazur:** No he doesn't consider this convergent evolution. What he's saying is that the cell is not able to put together protons, neutrons, and electrons to build a zinc, a cobalt, a magnesium or an iron atom. He says these atoms come from the external environment and that "it is not surprising that the functions and structures of the cell were obliged to follow and perpetuate the periodicity inherent in the chemical elements and the minerals." So that periodicity in biology, similar patterns that arise in different animals and plants, this sudden emergence of functions in mammals that first occurred in earlier invertebrates, for example, may be "directly correlated with periodicity at the level of the chemical elements".

**Lynn Margulis:** I haven't followed his papers in recent years, but I'm delighted to hear that he's working and publishing at 87.

About minerals, no doubt minerals are deeply involved in evolution. No doubt. At least 20 elements of the periodic table are absolutely and uniquely required for life today, according to Bob Williams, Oxford University Professor of Chemistry.

He showed that in addition to "CHNOPS" (carbon, hydrogen, nitrogen, oxygen, phosphorus, sulfur) -- now up to 20 elements are essential to living organisms today. Earlier the number probably was lower. These elements can't be substituted for each other.

If manganese is needed, magnesium won't work. And if magnesium is required for a certain enzyme, selenium won't substitute.

And the thing about the crystals of the elements, their mineralogical features, is that unlike components in biology when a pure substance is made of those chemical elements – there is identity. Elements are as identical as anything can be; this is not true of cells.

**Suzan Mazur:** Do you find the concept of morphogenetic fields interesting, which Scott Gilbert and Rudolf Raff have given some thought to, as well as Stuart Pivar with his focus on "biological structuralism" based on evolutionary changes of the torus? And Mark McMenamin's recent description in *Paleotorus* of the Seilacher tongue fossil as a flattened metazoan torus (convergence of longitudinal lines near the poles), an Ediacaran toroid? Mark McMenamin says this is an example of morphogenetic fields.

**Lynn Margulis:** The person who does promote the "morphogenetic fields" concept is Rupert Sheldrake, of course.

I haven't had a chance to look at the *Paleotorus* manuscript. I'm unfamiliar with torus thinking. But I'm very familiar with Mark McMenamin and his work. I've been in the field with him, he is a co-author. His analysis of Ediacara biota "metacellularity" is excellent. It is a kind of multicellularity but it's not ***"metazoan metacellularity"***. I think he's probably right that Ediacaran metacellularity differs from both plants and the familiar animal multicellularity. Animal tissue cells differentiate specific structures that include desmosomes, septate junctions, synapses, tight junctions and others. Plant cells are connected by plasmodesmata. I suspect the Ediacara biota comprised mostly syncitial protoctists that, in any case, as McMenamin and Dolf Seilacher show, were not animals.

**Suzan Mazur:** I believe you've made the statement that the "neo-Darwinists are a minor twentieth-century", now twenty-first century, "religious sect within the sprawling religious persuasion of Anglo-Saxon Biology."

**Lynn Margulis:** I did.

**Suzan Mazur:** Richard Dawkins gave a speech last year at the School of Ethical Culture in Manhattan where he was questioned from the audience about his embracing Darwinism as a religion And he said "I'm guilty" and I will make an effort to "reform".

**Lynn Margulis:** Good! Let me mention a marvelous book by Ronald W. Clark called *The Survival of Charles Darwin*. The first half or so of the book is Darwin, Darwinism and Darwin in his day. The second half is neo-Darwinism as it happened after

Darwin's death. It describes beautifully the political and social pressures on neo-Darwinism.

The story is connected to Mendelian genetics. Gregor Mendel was no monk in some very quiet garden in Brno in what is now the Czech Republic. He was a well-educated, fine naturalist who stayed in touch with the pope. He was aware of Darwin's views of change. Mendel's concepts were marvelous, scientifically impeccable.

I'll give you a simplified example comparable to Mendel's sweet pea botanical breeding work. Red flowers and white flowers breed true generation after generation. Red have red offspring. White crossed with white have all white offspring. But a hybrid cross of the breeding red with the breeding white flowers yields all pink flowers. When you cross the pink ones amongst each other, the offspring flowers are just as red or just as white as the parents or the grandparents.

The point is Mendel showed that the changes in heredity are mixtures, but they're pure. Pink parents produce red, white and pink offspring like parents or grandparents. No evolution of novelty occurs at all because exactly the same hereditary "factors" begun with are transmitted and remain pure through all subsequent generations. Mendel envisaged "factors" (now "genes") that do not change through time.

Mendel rejected the evolution stuff in the air at his time as bunk. He states the laws of heredity yes, but none leads to change or speciation.

Then Darwin writes that if there are no "gradual changes through time", his theory is wrong. So whatever else Darwin said, he certainly made it very clear that there were changes in life through time and those changes can be inherited. The basic point was the perceived contradiction between Mendel's claim of heritable total stability and Darwin's description of change. Only the inherited changes, he wrote, "are of interest to us".

What I think is that both Mendel and Darwin, except for Darwin's insistence on gradualism at the fossil record level, were correct. The changes through time are absolutely documented. The sciences of paleontology and paleobiology document change and common ancestry through time. But in detail those changes tend to be more discontinuous, saltatory ("punctuated equilibrium") than Darwin's great expectations of them.

**Then some smart-ass Cambridge mathematicians – cocktail-party types – J.B.S. Haldane, who was brilliant beyond words, and R.A. Fisher, an algebraicist as well, tried to reconcile the stability of Mendel's inheritance "factors" (which Johannsen later called "genes") with Darwin's insistence on change.**

Darwin was a geologist. He tried to trace lineages through time. He observed examples of extinction and appearance of life forms in the fossil record. Much fossil life is not like that of today but it's close enough so that we know that the extinct organisms were directly related to descendants of today, *e.g.*, that dinosaurs were related to today's reptiles, *etc.*

**And what Haldane, Fisher, Sewell Wright, Hardy, Weinberg *et al.* did was invent. They took genetics and made up "population genetics", based on the extrapolation of Mendel's rules. The superstructure is a theoretical one that ran away with the entire so-called "field". The neo-Darwinism after Darwin's death claimed to have resolved the Darwin (change)-Mendel (no change) contradiction.**

**X-rays were then revealed by Hermann J. Müller to cause mutation. Those genes do change. Any organism was envisaged to be made of its alphabet of genes: A, B, c, d, E, F, G, h, I, J, k . . . all the way to Z. (The A-to-Z-total is the**

**organism's genome.) "Big A" mutates back to "little a" and "little a" changes back to "big A" in a reciprocal fashion at a measurable rate. The notion is that if we accumulate enough gene change, enough genetic mutations, we explain the passage from one species to another. This is depicted as two branches in a family tree that emerge from one common ancestor to the two descendants. An entire Anglophone academic tradition of purported evolutionary description was developed quantified, computerized based on what I think is a conceptual topological error.**

**Suzan Mazur:** How are you doing at Oxford?

**Lynn Margulis:** I'm thriving, I have found excellent colleagues and friends in this place of the three "B's": biophilia, bibliophilia and bicyclophilia. However, I admit I feel burdened by neo-Darwinist tradition that still prevails among so many that has little to do with me, and what I feel compelled to ignore as useless for my own research.

**The Anglophone tradition was taught. I was taught and so were my contemporaries. And so were the younger scientists. Evolution was defined as "changes in gene frequencies in natural populations". The accumulation of genetic mutations were touted to be enough to change one species into another.**

**Furthermore, it was admitted, from the very beginning because it was measurable, that more than 99% of all detectable mutations, heritable changes were negative, mutations were mainly deleterious. They rationalized. One sees that less than 1% of genetic mutations, measurable heritable change, are not deleterious. They are presumably favorable. If enough favorable mutations occur, was the erroneous extrapolation, a change from one species to another would concurrently occur.**

**Suzan Mazur: So a certain dishonesty set in?**

**Lynn Margulis: No. It was not dishonesty. I think it was wish-fulfillment and social momentum. Assumptions, made but not verified, were taught as fact.**

**Suzan Mazur: But a whole industry grew up.**

**Lynn Margulis: Yes, but people are always more loyal to their tribal group than to any abstract notion of "truth" – scientists especially. If not they are unemployable. It is professional suicide to continually contradict one's teachers or social leaders.**

**Suzan Mazur:** Frank Turner, director of the Beinecke Rare Book and Manuscript Library at Yale has told me that scientists are "the most successful intellectuals in securing public funds" and in exchange for government grants agree to work to "ensure better health, economic stability and national security". Do you agree with that description and how do you feel about scientists in effect having to take a pledge to ensure national security in order to get funding for their labs?

**Lynn Margulis:** We did research that's directly related to the anthrax bacteria. First, you're talking here about American science. And secondly, you probably mean NSF, and even worse DARPA-funded science.

Scientists in general need funds. An aspect of science that is less true of the humanities or arts, is that they must chase money because the rate of cash flow must increase to increase research activity. Scientists almost always need equipment, materials, man-power, in short, money. All except some theoretical scientists need money and, for research results to be accurate and meaningful even the "theoretikers" must work with experimentalists who always need money.

So the economic system to me is such that university people, like most everyone else, maximize the rate of cash flow per square foot of institutional space. That is the main pressure. Scientists, like anyone else, follow the money flow. Many are entirely honest about it. Some of them will make bombs. Most won't go that far. The humanities and philosophy scholars receive far less public and corporate money because, in general, what they do is not perceived as practical. All they do is make books and teach esoterica to students.

**Suzan Mazur: But in making this agreement to ensure better health, economic stability and national security – does this affect the science? Does it constrain science? Is this one of the problems we have, why we're stuck in old science regarding evolution? Is there a certain "cycle of submission" at play here?**

**Lynn Margulis: Yes, some aspect of this prevails But evolution in addition has peculiar problems. Yes, science is traditional But I think it does have a self-correcting feature to it – people die basically. It takes a few generations to self-correct.**

**Suzan Mazur:** Is it too soon for an international gathering of holistic evolutionary science where mineralogists powwow with cell biologists, *etc.* and the conference is streamed over the Internet? Or is there still not enough of a common language?

**Lynn Margulis:** Yes there is some common scientific language but only among a limited number of people. Some have already met. Do you know Lisdisfarne's William Irwin Thompson?

**Suzan Mazur:** No I don't.

**Lynn Margulis:** He's an honest guy, profoundly well-educated, he is a scholar who founded the Lindisfarne Fellowship that for 25 years or so was funded mainly by Laurance A. Rockefeller, who had great confidence in Thompson. He admires science and thinks broadly, like a scientist, but enjoys many other intellectual concerns. He tends to learn science although he's not a scientist himself. He's just a fabulous intellectual. Basically he's a cultural critic.

**Suzan Mazur:** But we haven't had a conference where all these people come together and it's streamed on the Internet. We've had conferences with books, CDs and videos generated where most of the public is not let in.

**And mainstream media is retarded when it covers evolutionary science.**

**Lynn Margulis: Individual people in the mainstream media try to report comprehensively but are stopped because the mainstream media won't publish what it doesn't like or understand. Also, often the excellent scientists can't explain themselves in a way other people understand.**

**Suzan Mazur: Should the language be broken down a little like legal jargon has been simplified?**

**Lynn Margulis: Let's finish our conversation about the components of evolution and with what aspect I disagree. So much of evolution can not be disagreed with by someone who calls himself a scientist One component is natural selection. Natural selection occurs.**

**Suzan Mazur: Lima-de-Faria's definition of evolution includes the nonbiological and does not include natural selection. Mineralogist Bob Hazen has said that we need to define these terms. People are not in agreement about what evolution or selection is.**

**Why should Bob Hazen be left out of the conversation because he's a mineralogist?**

**Lynn Margulis: He isn't left out.**

**Suzan Mazur: Well he is in the sense that his friend Niles Eldredge told me: "He's [Bob Hazen's] a mineralogist. He's not an evolutionary biologist. So be careful."**

**Lynn Margulis:** Well Niles Eldredge, a wonderful friend and colleague of mine, is talking about those scientists who derive from zoology. He probably refers to the deliberate intellectual activity that reconciles Mendelian stability with Darwinian gradual change and tries to force it into this procrustean population genetics neo-Darwinism.

Francisco Ayala is presenting at the "evolutionary mechanisms session" in Rome. He was trained in Catholicism, Spanish-style, as a Dominican. We were in California at a meeting with Whiteheadian philosopher John Cobb. **At that meeting Ayala agreed with me when I stated that this doctrinaire neo-Darwinism is dead. He was a practitioner of neo-Darwinism but advances in molecular genetics, evolution, ecology, biochemistry, and other news had led him to agree that neo-Darwinism's now dead.**

The components of evolution (I don't think any scientist disagrees) that exist because there's so much data for them are: (1) the tendency for exponential growth of all populations – that is growth beyond a finite world; and (2) since the environment can't sustain them, there's an elimination process of natural selection.

The point of contention in science is here: (3) Where does novelty that's heritable come from? What is the source of evolutionary innovation? Especially positive inherited innovation, where does it come from?

It is here that the neo-Darwinist knee-jerk reaction kicks in. "By random mutations that accumulate so much that you have a new lineage." This final contention, their mistake in my view, is really the basis of nearly all our disagreement.

Everybody agrees: Heritable variation exists, it can be measured. Everybody agrees, as Darwin said, it's heritable variation "that's important to us" because variation is inherited. Everyone agrees "descent with modification" can be demonstrated. And furthermore, because of molecular biology, everybody agrees that all life on Earth today is related through common ancestry, as Darwin showed.

Everybody agrees with ultimate common ancestry of Earth's life, because the DNA, RNA messenger, transfer RNA, membrane-bounded cell constituents (lipids, the phospholipids) that we share – they're all virtually identical in all life today, it's all one single lineage. So that part of Darwinism – that we're all related by common ancestry – no scientist disagrees with.

The real disagreement about what the neo-Darwinists tout, for which there's very little evidence, if any, is that random mutations accumulate and when they accumulate enough, new species originate. The source of purposeful inherited novelty in evolution, the underlying reason the new species appear, is not random mutation rather it is symbiogenesis, the acquisition of foreign genomes.

When Salthe says we haven't seen that, he's talking about new species. He's not saying we haven't seen natural selection, he's saying we haven't seen natural selection produce new species, this particular aspect of neo-Darwinism.

**Suzan Mazur:** Were you invited to the Altenberg event? The Extended Synthesis meeting in July in Altenberg, Austria?

**Lynn Margulis:** No. I don't know anybody who was.

**Suzan Mazur:** Stuart Newman.

**Lynn Margulis:** I don't know Stuart Newman.

**Suzan Mazur:** His theory of form based on a pattern language he calls dynamical patterning modules was the centerpiece at Altenberg.

He proposes that all 35 or so animal phyla physically self-organized from single-celled organisms without a genetic recipe [i.e., genetic program] by the time of the Cambrian explosion half a billion years ago using this pattern language of DPMs. Natural selection supposedly followed.

**Lynn Margulis:** But they all use the word "multicellular" when they really mean "animal"– since there are no unicellular animals, and since multicellularity, genuine multicellularity, details of multicellularity, are known in protoctists, bacteria and all the major groups of life. He can't be correct in his claim of a lack of "genetic recipe" in life prior to animals. Every single cell lineage has DNA genes and has generated multicellular descendants. Some bacteria like *Gomphosphaeria* are always multicellular in all stages of development.

Many are unaware of this. Zoologists often display their dangerous ignorance; they play with 1/5$^{th}$ of the deck in biology. They belong to a thought style I don't share.

**The problem is that many fine scientists recognize genuine difficulties with the "standard model" of evolution, so to speak. However, most lack conceptual tools to solve the difficulties they legitimately recognize. They think they understand processes, like speciation [which in fact, is due mostly to karyotypic fissioning, symbiogenesis and cross-species hybridogenesis (as larval transfer)]. All of these are anastomoses. They are mergers of ancestral lineages. They can not realize that "development" is really microbial ecology and community successional change. These are entire traditions dismissed as "of historical interest only", fields "modern scientists" tend to know little or nothing about.**

**Suzan Mazur:** Do you have any comment on various numbers of genes humans are reported to have?

**Lynn Margulis:** What is the current number? 25,000?

**Suzan Mazur:** Well they're saying 20,000 - 25,000 up to 30,000. Then I interviewed industrialist and philanthropist David Koch, who's on the board at MIT, and he said he sat in on a meeting with some Nobel Prize winners there and in their discussion of genes the numbers ranged from 15,000 to 30,000.

**Lynn Margulis:** That is only a factor of two, nearly nothing. When they are off by factors of a million ($10^{5th}$ or $10^{6th}$) or, for example, off by $10^{20th}$, then it is serious, it is time to examine the inevitable unstated assumption or unjustified extrapolations.

**Suzan Mazur:** But they don't know what the gene really is at this point or what its origin is.

**Lynn Margulis:** I wouldn't say that. Many know the genes as nucleotide sequences in DNA. But they have pre-conceived notions about "programmed organisms determined by DNA" and "accumulation of random mutations", etc. that interfere with learning. These are scientific beliefs that, in their preconceptions, are like all the rest of the religions to which people are prone.

**I have not explicitly told you what I think is the major source of novelty in evolution, *i.e.*, what heritable genetic variation does lead to speciation. If, as I claim, heritable variation mostly does NOT come from gradual accumulation of random mutation, what does generate Darwin's variation upon which his Natural Selection can act? A fine scientific literature on this theme actually exists and grows every day but unfortunately it is scattered, poorly understood and neglected nearly entirely by the money-powerful, the publicity mongers of science and the media. Worse, much of it is not written in English or well-indexed. This literature shows that symbiogenesis, interspecific fusions (hybridogenesis, gene transfers of various types, karyotypic fissioning, and other forms of acquisition of "foreign**

**genomes" or epigenesis) are more important than the slow gradual accumulation of mutation or sexual mergers. If you are interested at all in this literature start with our *Sciencewriters* book *Acquiring Genomes: A Theory of the Origins of Species* (Margulis & Sagan, Basic Books, NY paperback).**

**Suzan Mazur:** I know you don't make an issue of being a woman in science and your concept of symbiogenesis is rooted in your predecessors who you think got it right (like Boris Mikhailovich Kozo-Polyansky's 1924 analysis soon to appear in English translation), but would you say the idea of evolution by cooperation is still largely something difficult for men of science to embrace and that may be why the "survival of the fittest crowd" are still so dug in?

**Lynn Margulis: The problem is NOT "competition versus cooperation". Those words are totally inappropriate for life. The language of life is metabolic chemistry. Even bankers and sports teams have to cooperate in order to compete. It's crucial to realize that it doesn't matter what team you're on, when you compete, even in sports where the term is valid, you still cooperate!**

## Chapter 19

# Paul Nurse: Revolution In Biology

*April 11, 2009*
*12:26 pm NZ*

This story first appeared on *CounterPunch*
http://www.counterpunch.org/mazur04102009.html

> *"I have an idealistic view of science as a liberalising and progressive force for humanity. . . . It is also a truly international activity which breaks down barriers between peoples of the world, an objective that always has been necessary and never more so than now."*
> – **Paul Nurse,** *Le Prix Nobel*

The elevator was supposed to go down to the powder room but mysteriously went up to the 8th floor at Manhattan's Rockefeller University, where I was attending a two-day public Evolution symposium last May. As the doors opened, in walked Nobel Prize-winning molecular biologist and Rockefeller University president Paul Nurse. I had emailed him, coincidentally, about an interview, and told him so while riding back down with him to the main floor. He was engaging, if perhaps a bit startled by the encounter (I'm taller). Once outside the building he made a quick escape and was soon halfway up the walkway to the talks at the Buckminster Fuller geodesic dome calling "Send me your stories!".

We bumped into one another again that night at the Evolution cocktail party and spoke briefly about the *Charlie Rose Show* science roundtables he'd co-hosted. I recontacted him recently and he agreed to a phone interview.

I didn't know a lot about the man Paul Nurse at the time of the Rockefeller event beyond the extraordinary charm he exudes on television and his public honors and accomplishments. He was knighted by Queen Elizabeth in 1999 for his excellent cancer research, awarded the Nobel Prize in 2001 for discovering key regulators of the cell cycle, given the French Legion d' Honneur in 2002, and named president of Rockefeller University in 2003. He's also a fellow of the Royal Society, a foreign associate of the National Academy of Sciences and foreign honorary member of the American Academy of Arts and Sciences.

I was happy to discover that he is self-made. Nurse grew up in the English countryside. His mother was a cook and part-time cleaner, his father a chauffeur and handyman and later a mechanic for Heinz.

Nurse writes in his autobiography that he fell in love with nature on long walks to school and became a biologist because of this. He says he had a difficult time in secondary school memorizing and didn't test well. But he was so talented in science that the University of Birmingham eventually waived his French language admission requirement, for instance. Until it did, he worked for a lab associated with Guinness brewery.

Paul Nurse received his undergraduate degree in biology from the University of Birmingham and his PhD in molecular biology from the University of East Anglia. He was a professor of microbiology at Oxford for many years and also served as Director General, Imperial Cancer Research Center. His current research at Rockefeller University involves "the molecular machinery that drives cell division and controls cell shape".

But he says he is "a complete amateur" when it comes to things like complexity and evolution.

Our interview follows.

**Suzan Mazur**: I'd like to pick up on the point you made at the World Science Festival in response to Stanford University physicist and string theorist Leonard Susskind saying there are elements of biology and evolution in physics, and that in 1905 physics "moved out of the domain of ordinary experience . . . into the domain of high velocity."

To which you said:

"Maybe biology is on the edge of something similar to 1905 physics with the emerging complexity of biological systems – in fact, a move from straight forward linear causality. And I do wonder whether biology may go through a revolution in the coming decades."

Can you expand on that?

**Paul Nurse**: Well I can expand a little bit, not enormously, as follows.

What I think I'm thinking about is that biology deals with complexity, with networks that are linked together in very complicated ways with all sorts of feedback, positive and negative. All sorts of redundancies, that is, if one part of the network fails, there are other ways of compensating for it.

Furthermore, networks that can change, unlike normal computer hardware. So you could rewire the networks in different ways in space and time. In cells and organisms. And as a consequence of this, a very complex set of behaviors can emerge.

They could be on the edge of chaos, which has been discussed a lot by others. They are difficult to predict intuitively.

**Suzan Mazur**: A number of biologists knowledgeable in the physical sciences have proposed that multicellular form and pattern are as much a function of the physics of mesoscale materials as of the gene products that constitute those materials. This has been claimed to present a direct challenge to Darwinian scenarios for the origin of organismal body plans. I was wondering if you see this as a potential 1905 moment for biology.

**Paul Nurse**: I don't agree with the challenge to "Darwinian scenarios".

Darwin himself argued that although natural selection is a driving force for change, there are restraints on organisms and how and where they can evolve. And that restraint, as he argued in 1859, is really the equivalent of what you refer to.

**Suzan Mazur**: Are you saying that "restraint" that Darwin referred to is the equivalent of the evolutionary mechanisms these biologists are referring to?

**Paul Nurse**: What they, I think, are referring to is the fact that living things can't occupy all of the phase-space. That is to say, there are certain limited stable states and conditions.

That's one that's been argued on and off for many years. I remember when I was a graduate student, for example, Brian Goodwin used to argue this. Darwin himself argued that there are restraints. Obviously, living things made of proteins and nucleic acids, denature at high temperature. These living organisms can't live at very high temperatures. They have to live in watery environments at reasonable temperatures.

Those are the sort of restraints I think probably that Darwin was thinking of.

**Suzan Mazur**: Do you think the emerging complexity of biology might require a new mathematics or is it the reverse that the mathematics of straightforward linear causality is inappropriate in the first place and needs to be replaced by math for which the emergent complexity of life is an epiphenomenon of more fundamental physical processes that these mathematics model?

**Paul Nurse**: It may require a new mathematics. Because I'm not a mathematician I'm not quite sure what that might generate/do.

It may require a different sort of language, by which I mean, quite often what biologists do is make interaction maps. Does A touch B touch C touch D and so on.

But, in fact, the nature of those interactions varies. Sometimes they just touch and do nothing. Sometimes they touch and turn into something else. Sometimes they touch and change another connection.

Using simple network analogies, like transport networks for example – it's not appropriate because it's not reflecting the biology. It's reflecting a man-made simplified interaction network. So we may need different language which could lead to different mathematics to describe this – and that this is not going to be intuitive, to go back to my earlier point.

We perhaps have to think, I've sometimes argued this, of better ways to move from the chemistry of life, which we're rather good at describing, into how that chemistry is translated into the management of information.

There are very good examples of that. DNA is a chemical that has a structure that can be translated into a digital information storage device. And we understand its role in biology in terms of a digital information device.

What we think about is the storage of information and how it's translated. I think we may need to think of better ways of translating the chemistry, which is what most of us do, into the management of information.

**Suzan Mazur**: This is a bit off the point, but I was just watching a 1986 television interview with Isaac Asimov, where he said "I have no hesitation in invading fields." He said he knew nothing about astronomy, for instance, before he started writing his books. That he was self-educated in astronomy. He thought there was no harm in generalists in science and that the danger was too much specialization.

I wondered if you had any thoughts about that. The fact that you're presenting speakers at the World Science Festival to a public audience and on the *Charlie Rose Show*. You must think that it's important that the wider audience be brought into the thinking and discussion of science and evolution.

**Paul Nurse**: I think it's very important for science and scientists to be talking to the general public.

**Suzan Mazur**: And on the evolution issue, particularly?

**Paul Nurse**: For example, evolution. But actually on many topics, simply because we receive enormous support from the public and the public has to think that there is something there that's worth supporting, which is, improving the health, wealth creation and better quality of life, etc.

But it's probably a different point you're making, which is a more generalist one I guess.

**Suzan Mazur**: Well there are some ideas out there in the public that could be useful.

**Paul Nurse**: I don't know. When we say the public, I certainly think that there are ideas out there which touch lots of different disciplines. So I think having interdisciplinary, multi-disciplinary approaches could be interesting. Whether they will come from the public, I'm not sure.

**Suzan Mazur**: That's what's happening now because of the Internet.

**Paul Nurse**: It is but there's so much garbage on the Internet. And everybody wants to have a grand idea.

Like you were saying, people say, oh they've proven Darwin wrong. I mean this is just publicity really. It isn't real sort of argument.

**Suzan Mazur**: So you're saying Darwin has already said this.

**Paul Nurse**: In that case yes, but it's all become too much of a totem. Of course, how could Darwin get everything right in 1859? I mean it's just ridiculous. It's just one of these really stupid arguments.

**Suzan Mazur**: New language to say the same thing?

**Paul Nurse**: Well the new language part was really just to try and deal with the complexity of nature. And I'm saying nothing new. People have talked and discussed about this for ages.

Kant was the first person I came across talking about systems, etc. over 200 years ago. But more recently since the Second World War, in the 1940s and 1950s information theory, game theory, Shanon, Weinberg, *et al.* – these characters thought a lot about information, and how it's managed.

This was overshadowed by the molecular biology revolution, which has a very powerful way of understanding things. It didn't focus on complexity and indeed could not encompass that complexity. And now I see what's very interesting is that we can use the rigor and tools of molecular biology and molecular genetics, combined with a better understanding of complexity.

That's much more fertile ground for these discussions than we've had for 30, 40 years.

**Suzan Mazur**: Your own work on evolution at present is on what?

**Paul Nurse**: Everything I'm talking about here, both in evolution or in terms of complexity is not my research area – I'm a complete amateur.

**Suzan Mazur**: What panel are you presenting at the World Science Festival this year?

**Paul Nurse**: They've just written to me about this. I think I'm doing something on storytelling.

# Appendix A:
# Stuart Kauffman: Rethink Evolution, Self-Organization is Real

*May 5, 2008*
*1:21 pm NZ*

**Suzan Mazur**: You were one of the pioneers of self-organization. I've looked at your new book, *Reinventing the Sacred*. You're thinking in a much bigger way.

**Stuart Kauffman**: It has to do with getting older. You don't write a book called *Reinventing the Sacred* when you're 30...

**Suzan Mazur**: Are there alternatives to natural selection?

**Stuart Kauffman**: I think self-organization is part of an alternative to natural selection. Let me try to frame it for you. In fact, it's a huge debate. The truth is that we don't know how to think about it.

**Suzan Mazur**: You said in your forward to *Investigations*: "Self organization mingles with natural selection in barely understood ways to yield the magnificence of our teeming biosphere. We must, therefore, expand evolutionary theory."

**Stuart Kauffman**: I'm still there... *Investigations* is the weirdest book I've ever written and it is the prelude to *Reinventing the Sacred*. I've gone through a trajectory in my life. I started with pure self-organization and I originally thought 40 years ago –

let's see how far we can get without any selection at all... Just think about a snowflake.

**Suzan Mazur**: You've said: "The snowflake's delicate six-fold symmetry tells us that order can arise without the benefit of natural selection." So it can arise without natural selection, but it's not living.

**Stuart Kauffman**: But it's not living. Right. There are all sorts of signatures of self-organization. I'll give you one that very few would doubt. I don't spend time talking about it in any of my books. But here it is.

If you take lipids like cholesterol and you put them in water, they fall into a structure – a liposome, which is called a bilipid membrane, that forms a hollow vesicle... Now if you look at the structure of this bilipid membrane, it's virtually identical to the bilipid membrane in your cells. So this is a self-organized property of lipids.

That's physics and chemistry... And evolution has made use of it to make lipid membranes that balance cells. So that's a snowflake. It's hard to look at that and doubt it. Nothing mysterious or mystical...

**Suzan Mazur**: No genes in the mix.

**Stuart Kauffman**: Genes by themselves are utterly dead. They're just DNA molecules. It takes a whole cell in the case of a fertilized egg to grow into an adult. So there's a lot of physics and chemistry...

And somehow the right answer is that this is a whole integrated system in which matter, energy, information, whatever that means – it turns out to be a very slippery concept – and the control of process is all organized in some way.

The philosopher Immanuel Kant talked about this – the self-propagating organization of process...

**Suzan Mazur**: You say in *Reinventing the Sacred*: "I have always believed that that basis of life is deeper and that it rests on catalysis. The speeding up of chemical reactions by enzymes."

**Stuart Kauffman**: And then I have a chapter called "The Cycle of Work".

**Suzan Mazur**: You say: "My second intuition is that it's based on some form of collective autocatalysis."

**Stuart Kauffman**: Right. So remember Charles Darwin starts with life. He doesn't get you to life...

**Suzan Mazur**: Are you saying form came first and genes later?

**Stuart Kauffman**: You mean in the origin of life.

**Suzan Mazur**: Yes.

**Stuart Kauffman**: I'll tell you what I think. Current cells use DNA, RNA and proteins. It's really unlikely that the earliest life on Earth used anything as complicated as contemporary DNA, RNA and protein, because the machinery by which our DNA gets translated into proteins is incredibly complicated and it includes the fact that the genes code for that protein that carries out the translation for those proteins. They're called amino acid synthesizers. So life couldn't have started out that complex.

Assuming life started on Earth – it had to start somehow else and evolve into current life. People are working on the origin of life, including me, including my idea on collective autocatalysis. It is a debate about self-organization. But it's before there is life.

There's a guy named Reza Ghadiri. And Reza has made a collectively autocatalytic system of proteins where protein 1 catalyzes the formation of protein 2 out of protein 2 parts and protein 2 catalyzes the formation of protein 1 out of protein 1 parts.

There's no molecule in Ghadiri's system. This system catalyzes its own formation. The set as a whole is collectively autocatalytic. It achieves catalytic closure. That's a done deal experimentally. Molecular application's in the bag. Ghadiri at Scripps Research Institute has done it.

Now before he did that he also made a protein that catalyzed its own formation. So that's both logically possible and that's in the bag experimentally too.

Next thing to tell you is that a cell really is a collectively autocatalytic whole. There is no molecule in the cell that catalyzes its own formation. The cell as a whole builds itself...

**Suzan Mazur**: Originally genes were or were not part of the story?

**Stuart Kauffman**: Nobody knows.

**Suzan Mazur**: Your sense is that it was more of a mechanical and chemical process first.

**Stuart Kauffman**: My sense is that it was a catalytic process. Collectively autocatalytic. I have a whole theory about it – chapter 5 in the new book. But that's just a theory.

What Reza's done is fact. Whether the theory turns out to be correct, we don't know. It's a beautiful theory.

**Suzan Mazur**: But in the beginning when you had this simple cell there were no genes.

**Stuart Kauffman**: It depends what you mean by genes. If you mean by a gene a sequence of nucleotides that codes for a protein, I think it's extremely unlikely that at the start of life – if life started on Earth or wherever it started – that you started with genes that coded for proteins. That's just utterly remote.

But it may have been that the earliest catalysts were polynucleotides rather than proteins or something else. In that sense of gene – yes. But they wouldn't have coded for proteins. It's just remote.

So what we're talking about is how do you get life in the first place?

**Suzan Mazur**: Where do the work cycles fit in?

**Stuart Kauffman**: Think about choo-choo trains. A train uses heat. It turns it into mechanical work train pistons. That's a work cycle. It uses the transfer of heat from a hot to a cold place. And it manages to get the pistons to go around.

It's been around since 1830. A guy named Sadi Carnot worked out the principles of a thermodynamic work cycle.

I think that an essential part of life is that it does work cycles. It's not enough that life is markedly reproducing... Every free living cell, in fact, all the cells in your body do work cycles – chemical work cycles and mechano-chemical work cycles. And that's missing from what most people think about life... Buried in this are the roles of self-organization and natural selection.

Selection couldn't have played a role before there were organisms. You couldn't have had natural selection because there were no organisms. It's a different debate whether some other form of selection for chemical stability might have played a role.

There are some physicists who are asking questions like: Is natural selection an expression of some more general process? Like entropy production. And it's all up in the air. But at least people are thinking about it. Meanwhile, we've got self-organization.

**Suzan Mazur**: Are evolution and development the same thing?

**Stuart Kauffman**: Sure.

**Suzan Mazur**: You mention Nobel laureate Murray Gell-Mann in your book. I know you were colleagues at the Santa Fe Institute. Do you think similarly about self-organization?

**Stuart Kauffman**: Probably not... Murray is a profound reductionist. He's been a major voice in the Santa Fe Institute and is a superb scientist. I'm not a reductionist. Reductionism meaning everything is due to the physical laws down there.

What's happening is that the physics community is dividing now...

But let me tell you where I started 40 years ago. And where it is now. I literally started on this when I was 24 and I'm 68... That was about 1964.

You know that cells get to be different from one another – cell differentiation. You make liver cells and kidney cells and spleen cells. And the question at the time was: So how do cells get to be different?

We thought – different cells get different genes from the fertilized egg. That turned out to be false. Just wrong. All the cells in your body have the same genes.

**Suzan Mazur**: Right.

**Stuart Kauffman**: Now it's essential to know that different cells in your body make different proteins. Red blood cells make hemoglobin. That's because different genes are active. Where active means making more protein.

Two guys who got the Nobel prize for this, Francois Jacob and Jacques Monod, in 1961 showed in bacteria that genes could turn one another on and off. This is absolutely essential now. One gene can make a protein that binds to a little DNA region near another gene and turn the other gene on or turn it off.

**Suzan Mazur**: Right.

**Stuart Kauffman**: So there's a sense – leaving out the rest of the physics and chemistry of the cell, which we cannot do, but just for the moment – then you could imagine genes turning one another on and off. Jacob and Minod published a document in which they said imagine you've got two genes.

You and I are the two genes. And we're both spontaneously active, if nothing happens to us. But Stu makes the Stu protein which goes over and binds next to the Suzan gene and shuts Suzan off. And vice versa. Suzan makes a protein that shuts Stu off.

So it's a tiny circuit, a genetic circuit. You can think of it like an electrical circuit. Then that circuit – and I think you can see this immediately – has two alternative steady states. Suzan on. Stu off. And Stu on. Suzan off. Can you see it?

**Suzan Mazur**: I think so.

**Stuart Kauffman**: So what they said was – look the same genome is giving rise to two patterns of gene activity. Suzan on. Stu off. And the other way around.

**Suzan Mazur**: Right.

**Stuart Kauffman**: This could be what controls cell differentiation. And they revolutionized the whole field of developmental biology with that paper.

I came along about a year later. And what I said was – we used to think 100,000 genes. We now know it's about 25,000 or 30,000 genes. And I thought, well, there's some sort of regulatory circuitry among these 25,000 or 30,000 genes. And there is. Forty-four years later we know something about it.

Imagine you've got 30,000 genes and somehow they're turning one another on and off in some complicated way. Okay. What I did – this is Stu's early foray into self-organization…

**Suzan Mazur**: So how many of the 25,000 or 30,000 are doing the turning on and off?

**Stuart Kauffman**: Nobody knew 40 years ago... Here's what we know now. In the human, there are approximately 2,000 genes that seem to play the role of turning one another on and off and the rest of the genes on and off... They're called transcription factors. And they're also regulating the other genes.

Hemoglobin is probably not regulating anything. It's regulated, but not regulating.

**Suzan Mazur**: Right.

**Stuart Kauffman**: So here's what I did. This is an essential core of current biology.

**Suzan Mazur**: The endogenous variables...

**Stuart Kauffman**: Right. You also have all the proteins... Let's suppose that there are 25,000 genes. And 2,000 of them are playing the role of regulating one another and regulating the other 22,500. Just imagine that genes can only be on or off. That's false. That's an idealization. Then how many possible patterns of gene activity are there?

Well there's 25,000 genes. So each could be on or off. So there's 2x2x2 25,000 times. Well that's $2^{25,000th}$. Right?

**Suzan Mazur**: Right.

**Stuart Kauffman**: Which is something like $10^{7,000th}$. Okay? There's only $10^{80th}$ particles in the whole universe. Are you stunned?

**Suzan Mazur**: It's getting pretty staggering...

**Stuart Kauffman**: So, 25,000 is plenty if you start thinking about all the possible combinations of their activities. It's super- hyper-astronomical.

**Suzan Mazur**: Right.

**Stuart Kauffman**: The next idea you need is somehow this network among the genes is controlling their activities. We don't know what this network is. My colleagues and I have just published a paper in which we think we maybe know. We have the first sketch of what this regulatory network looks like...

Anyway here's what I did when I was young. I asked the following radical question... I said does this regulatory network have to be really really special and tuned by natural selection to give rise to normal development? Or could it be spontaneously self-organized so that there's a huge set of possible networks and they're sort of all good enough? In other words, is it a spontaneous self-organized property of complex networks that they just do the right thing? ...

So I was saying ignore selection. Let's just ask whether or not there's a self-organized property and complex network of genes.

And what I showed in my mid 20s – I was 27 when I published it for the first time – was that my intuition was right. There really are. And so I modeled genes like they were lightbulbs, which they're not. And I made random lightbulb networks.

They're called Boolean networks because of a guy named George Boole. We now know a vast amount about the behavior of really complicated Boolean networks. Even random Boolean networks. So I'm just going to tell you a couple of things.

**Suzan Mazur**: Okay.

**Stuart Kauffman**: You know how I had you and me turning one another on and off and we had two steady states.

**Suzan Mazur**: Yes.

**Stuart Kauffman**: So the fancy word for those two steady states is attractor. That's the mathematical word. And you can think of it like a mountain region with a bunch of lakes in it. And each lake is like an attractor. And you know how streams flow into a lake. So in the space of all the possible pattern of gene activity, most of them constitute streams that flow into the attractor lakes. So the hypothesis I've had for 45 years, partially taken from Jacob and Monod, is that cell types – livers, kidneys, etc. – are these attractors.

**Suzan Mazur**: I see.

**Stuart Kauffman**: So one lake is a liver. Another lake is a kidney. Another lake is... You with me?

**Suzan Mazur**: Yes.

**Stuart Kauffman**: So we've got evidence that that hypothesis is true. Cell types look like they're attractors. Now, if that's true, cells getting to be different from one another happens in basically one of two ways.

You hop out of one lake into a mountain pass and flow down a creek into another lake. And then there's a fancier way in which you wiggle the mountains and change where the lakes are. That's called a bifurcation.

So this is sort of the two ways that it can happen. And we've got evidence for both. So we're beginning to understand that the cell and the organism is a very complicated set of processes activating and inhibiting one another. It's really much broader than genes.

**Suzan Mazur**: And form arises?

**Stuart Kauffman**: To say we know nothing about how form arises is wrong. There's been 70 years of superb developmental biology...

**Suzan Mazur**: Can you, for instance, do plastic surgery embryonically where the correction will take – say to an arm? ...

**Stuart Kauffman**: You mean could you conceivably take a thalidomide baby and do surgery and make it grow a normal arm?

**Suzan Mazur**: Yes.

**Stuart Kauffman**: Conceivably.

**Suzan Mazur**: You can?

**Stuart Kauffman**: No. Nobody's ever done it. But it doesn't seem impossible. And this has to do with what I'm working on right now. A lot of people are working on controlling and steering cell fates. That's exactly what I'm doing right now. I'm trying to get cancer cells to differentiate into normal cells. I'm trying to get a new way to treat cancer...

But to get back to self-organization, I showed two main things. Years ago I showed lakes of the kind you would need to explain cell types as lakes as attractors. And we know that cell types are actually attractors. It's early evidence. I think that it's very likely that it's true...

Now I'm going to tell you something that's just stunning. All of this work that has been done on random Boolean knots – it turns out that they can behave in three broad ways: ordered, chaotic, and there's a phase transition between the ordered regime and the chaotic regime where cells are poised at the "edge of chaos."

That's a phrase we came up with at Santa Fe Institute. A whole bunch of us – Chris Langdon, Norman Packard and I are the three main people who focused on all of this.

I have ever since 1987 believed that cells are poised on the edge of chaos. You'll find it in my first two books.

The easiest book of mine to read, by the way, is my second book: *At Home in the Universe,* which a lot of people have read. Al Gore read it. I wrote it with Gore in mind…

So there's this poised edge of chaos state between order and chaos. Here's what we're beginning to know now, 20 years later. There's evidence that cells are at the edge of chaos. The mathematical term is critical. Ordered, chaotic and critical. Edge of chaos will do.

So two main papers have been published. One came out just a couple of weeks ago and I'm one of the authors on it… It is the first direct evidence that maybe cells are at the edge of chaos. There's really dramatic evidence. It's gorgeous. But it's only one example.

**Suzan Mazur**: Can we draw conclusions?

**Stuart Kauffman**: No. But could you say – neat, let's explore it further? Yes.

**Suzan Mazur**: If it's right?

**Stuart Kauffman**: I suspect, I hypothesize that we may have found something general about life anywhere in the universe. That cells or whatever the analog of cells are anywhere are going to have to be at the edge of chaos because they could do all sorts of neat things.

They can coordinate the most complicated behavior. They can propagate information most efficiently. There are all sorts of neat reasons why it's incredibly advantageous to be at the edge of chaos.

Notice that I just used the word advantageous. Now you start hearing natural selection creep in. So it turns out that to be at the edge of chaos, networks have to be pretty special. They can't just be any old network. They have to be tuned to be at the edge of chaos.

And what could possibly be doing that tuning? Well, natural selection, because it's highly advantageous. So here is a marriage of self-organization and selection. Both are necessary.

In other words, the self-organization part is that large classes of networks have a property that they're either ordered, chaotic or edge of chaos – critical... So self-organization affords the capacity to be critical and then selection gets it and maintains it. And maybe it's so general that it's a law for any biosphere.

**Suzan Mazur**: So natural selection exists throughout the universe?

**Stuart Kauffman**: Well, yes, wherever there's life. But notice that there's self-organization too...

**There are people who are spouting off as if we know the answer. We don't know the answer.**

**Suzan Mazur**: So you're saying we should enjoy life.

**Stuart Kauffman**: Well, we should enjoy life. But we have to rethink evolutionary theory. It's not just natural selection. Self-organization is real.

# Appendix B

# Stuart Newman's "High Tea" Before Genetic Programs There Were DPMs

*March 31, 2008*
*9:14 am NZ*

While some scientists prefer shaping their opinions about evolution based on audience reaction while on book tour – others are actually looking for answers in the lab. Stuart Newman, Professor of Cell Biology and Anatomy at New York Medical College, is the real deal.

Newman will be presenting his full theory about "dynamic patterning modules" (DPMs) and "form" in evolution at a symposium in Altenberg, Austria this July at Konrad Lorenz Institute. The conference – first highlighted in my story "Altenberg! The Woodstock of Evolution?", carried by *Scoop Media* – is designed to discuss a remix of the existing theory of evolution.

Charles Darwin's theory was last updated 70 years ago. Extended Evolutionary Synthesis is the working title of the new one.

In a phone interview, Stuart Newman told me that some of his work on the theory of form – which the Modern Synthesis lacks – was done in collaboration with Gerd Müller, a theoretical biologist at the University of Vienna. Müller is also chairman of KLI and principal organizer of the Altenberg symposium.

Newman has co-authored the textbook *Biological Physics of the Developing Embryo* (Cambridge Univ. Press) with Gabor Forgacs, a biological physicist at the University of Missouri, and co-edited the volume *Origination of Organismal Form* (MIT Press) with Gerd Müller, also contributing a few chapters to it.

On Tuesday, March 25, Newman spoke at the University of Notre Dame's Interdisciplinary Center for the Study of Biocomplexity. Newman states that "in contrast to the Neo-Darwinian principle… phenotypic disparity early in evolution occurred in advance of, rather than closely tracked, genotypic change."

**Here's the abstract from Newman's Notre Dame "high tea":**

"The shapes and forms of multicellular organisms arise by generation of new cell states and types and changes in the numbers and rearrangements of the various kinds of cells. This talk will consider the role played by a core set of "dynamic patterning modules" (DPMs) in the origination, development and evolution of complex organisms. DPMs consist of the gene products of what is known as the "developmental-genetic toolkit," but considered in subsets, as dynamical networks embodying physical processes characteristic of chemically and mechanically excitable meso- to macroscopic systems like cell aggregates: cohesion, viscoelasticity, diffusion, and spatio-temporal heterogeneity based on lateral inhibition, and multistable and oscillatory dynamics.

I will focus on the emergence of the multicellular animals (metazoa), and show how the toolkit gene products and pathways that pre-existed this form of life acquired novel morphogenetic functions simply by virtue of the change in scale and context inherent to multicellularity. We show that DPMs, acting singly and in combination with each other, constitute a "pattern language" capable of generating all metazoan body plans and organ forms.

This concept implies that the multicellular organisms of the late Precambrian – early Cambrian were phenotypically highly plastic, fluently exploring morphospace in a fashion decoupled from both genotypic change and adaptation. The stable developmental trajectories and morphological phenotypes of modern animals, then, are considered to be products of stabilizing selection. This perspective provides a solution to the apparent "molecular homology-analogy paradox," whereby divergent modern animal types utilize the same molecular toolkit during development by proposing, in contrast to the Neo-Darwinian principle, that phenotypic disparity early in evolution occurred in advance of, rather than closely tracked, genotypic change."

# Appendix C

# The Enlightening Ramray Bhat: Origin of Body Plans

*April 15, 2008*
*12:32 pm NZ*

"Our theory does not stand against natural selection in its entirety – it relegates it to a less important role," says Ramray Bhat, cell biologist Stuart Newman's co-author of the just published paper in *Physical Biology*: "Dynamical patterning modules: physico-genetic determinants of morphological development". Nevertheless, neo-Darwinians – to whom natural selection is central to evolution – have tended to bury their heads in the sand when presented with a theory for form, which the neo-Darwinian model lacks. Why spoil next year's commercial celebration of the 150th anniversary of Charles Darwin's *Origin of Species* for a really coherent new theory?

I asked Ramray Bhat, now a graduate student at New York Medical College, if he'd answer a few questions about his paper and tell me about his journey from Calcutta to Valhalla and a collaboration with Stuart Newman. Our conversation follows.

**Suzan Mazur**: Are you and your distinguished co-author Stuart Newman saying that you have the first really coherent theory of evolution with regard to how virtually all of today's animal forms self-organized – 35 phyla – roughly a half billion years ago?

**Ramray Bhat**: I would rather put it as follows – we have the first really coherent framework to explain the origination and evolution of body plans and organ forms within a short evolutionary period, known as the Cambrian explosion.

Within this framework we also explore the relationships between gene products mediating the physics of biological matter (which we denote as DPMs) and transcription factors (DTFs), which carry out effects of the former within cells and tissues and hardwire DPMs' molecular players within regulatory networks. This framework also solves the Molecular Homology-Analogy paradox – why same/similar sets of genes are employed to build functionally or structurally similar organ forms in widely divergent organisms.

All of these are inconsistent with and cannot be explained by the classical neo-Darwinian model. We accommodate the role of natural selection in our framework mainly to lock the already-emerged but immensely plastic forms into place, and to render them robust.

**Suzan Mazur**: How do you define self-organization?

**Ramray Bhat**: By self-organization we mean the generic property of biological matter to attain a certain complexity in size, shape and pattern without depending on a blueprint or a recipe that is coded within its genome, or for that matter any other "ome". Rather this property comes from the physics and chemistry that make up biological matter.

**Suzan Mazur**: You state that you take a physicalist perspective. And you identify a "pattern language" that shaped multicellular life as it emerged from the single-cell state at the time of the Cambrian explosion. You call this language DPMs – dynamical patterning modules. The concept of DPMs is the main thrust of your paper just published online in *Physical Biology*, correct?

**Ramray Bhat**: Correct. While the paper published in *Physical Biology* has been written with emphasis on the physics and physical phenomena that DPMs embody, our contribution to the *International Journal of Developmental Biology* explores in some depth the relationship of DPMs with DTFs.

We also believe that the forms preceding the Cambrian explosion need not have been necessarily single-celled. They might have been transiently and variably multicellular sheets. Diverse three-dimensional "plastic" (*i.e.*, polymorphic) body plans came about within a short evolutionary period as a consequence of the DPMs acting singly or in combination with each other.

**Suzan Mazur**: You identify nine DPMs but say there are possibly more, as well as illustrate their effects. Are you saying in your paper that DPMs still exist? That while DPMs explored the formation of internal body cavities, segmentation, appendages, primitive hearts and eyes in highly plastic ancient multicellular organisms, DPMs continue to have a role in modern-day organisms – though only to a degree?

**Ramray Bhat**: Of course, DPMs continue to have a prominent role in the development of extant multicellular organisms. My experimental research involves trying to tease out the DPMs involved in patterning the skeleton of limbs. However, the DPMs can no longer explore as many new possibilities in terms of organ forms or body plans as they did in the ancient past; that is because their effects are now hardwired to the genome through millions of years of stabilizing evolution. The result of stable body plans and organ forms has come with a trade-off: the ability of the DPMs to freely explore what we call morphospace is severely constrained.

**Suzan Mazur**: Why did you decide not to present a model for DPMs?

**Ramray Bhat**: Our first task was to build a theoretical framework within which to resolve issues that loom large in the field of evo-devo, such as the Cambrian explosion and the Molecular Homology-Analogy paradox. This framework was built by assimilating previous research in this area by Prof Newman and his colleagues. Having put forward the same in the form of these two papers, we would embark on our next step shortly – to devise a computational model of the action of the DPMs.

**Suzan Mazur**: What do DPMs look like?

**Ramray Bhat**: DPMs *per se* are not readily visualizable like, say, proteins or genes, which are particular kinds of molecules. Each DPM consists of gene products and the physics they mobilize. For example, the DPM we annotate as ADH consists of cadherins and lectins and their associated physical property of adhesion. These DPMs were present in unicellular organisms but assumed their physical role (relevant to rapid and exhaustive exploration of multicellular form), only in the context of multicellularity. The effects of DPMs are easy to visualize, however, and we have done so in a series of "before" and "after" figures in our *Physical Biology* paper.

**Suzan Mazur**: You identify DTFs – developmental transcriptional factors – as coming into play as stabilizers but only after considerable body-building took place in the multicellular organisms at the time of the Cambrian explosion. Could you say a bit more about the role of DTFs?

**Ramray Bhat**: Developmental biologists have observed a small set of genes, coordinating organismal development, to be highly conserved across the multicellular kingdom. They call these genes the Developmental Genetic Toolkit. The genes whose products constitute DPMs (along with the physical phenomena they mediate individually, and in combination) are components of this toolkit.

There is another class of molecular players which also figure in the Toolkit but are not tied to any physics *per se*. They act in consequence to the action of DPMs to switch on and off certain genes and thus mediate cell-specific or tissue-specific effects of the DPMs. Since the DTFs are as ancient as the DPM Toolkit components, they had roles in the unicellular world in mediating transcriptional responses to internal and external signals.

Since embryonic regions and organs did not come into existence before multicellularity, the association of the DTFs with the DPMs as well as with their own biological effects can be regarded more as "frozen accidents". They may also have been the reason for a somewhat constrained genotype-phenotype relationship in extant organisms as they hardwire the DPMs by participating with them im regulatory networks.

**Suzan Mazur**: As a graduate student, does your experimental work in Stuart Newman's laboratory have anything to do with the DPM theory you have put forward in the two papers?

**Ramray Bhat**: An underlying theme in Prof. Newman's lab has been to not only study the molecular players involved in developmental phenomena and to tease out their dynamics, but also to understand the physics that goes along with their role. I investigate the pattern formation of limb skeletons – the underlying principles of which are ubiquitous in the vertebrate kingdom. A lot of the molecules and the dynamics involved in this process such as cadherins, galectins, morphogens and the Notch pathway molecules are components of the very DPMs we have described in the paper. Using a cell-culture model, I am trying to pin down a core network of molecules patterning the process of cartilage formation in chicken limb buds.

**Suzan Mazur**: Do you expect resistance to your theory in light of all the celebration surrounding the 150th anniversary next year of Darwin's publication of *The Origin of Species*?

**Ramray Bhat**: Science takes place through a continual process of dialectics. We expect resistance and debate from the scientific community regarding these new ideas, as it is only through this that a collective understanding of the origination and evolution of organismal form can be bettered. Our theory does not stand against natural selection in its entirety – it relegates it to a less important role – one of fine-tuning and building upon the body plans and organ forms, brought by actions and interactions of DPMs.

**Suzan Mazur**: How did you find your way from Calcutta to Valhalla and into Stuart Newman's lab and collaboration on a paper he has said reflects a synthesis of 20 years of his work?

**Ramray Bhat**: I developed an interest in evolution while in high school and studied organismal development and embryology during my first year in medicine at the University of Calcutta Medical College. This interest grew into a passion and alongside the drudgery of my medical curriculum I was largely teaching myself evo-devo through the works of Stephen Gould, Stuart Kauffman, Mary Jane West-Eberhard and Wallace Arthur.

During this period, I was immensely encouraged to learn more and guided to a great extent by prominent scientists in India such as Prof. Vidyanand Nanjundiah, Prof. Partha Majumder and Prof. Amitabh Joshi. It is through them that I came to know about the contributions in this field by Prof. Newman.

Prof. Newman's theoretical-cum-experimental approach to research on limb skeletal pattern formation, his formidable contribution to the growth of the evo-devo field, as well as his attention to the socio-cultural and philosophical aspects of biological research motivated me to apply to New York Medical College.

The inspiration to co-write this paper largely came about through continuous discussion with Prof. Newman as well as a long-held desire to resolve the above-mentioned evo-devo issues, which had been needling me ever since I started reading about evolution.

**Suzan Mazur**: How long have you been working with Stuart Newman?

**Ramray Bhat**: I have been working under him as a graduate student for the past two and a half years.

**Suzan Mazur**: How would you describe your collaboration with Stuart Newman?

**Ramray Bhat**: I would qualify my experience under him as more of an education than collaboration. Not only has my experimental research in his lab been stimulating, our daily, and sometimes into-the-night discussions on my research as well as biological theory and philosophy, which I feel is integral to an ideal scientific education, will go a long way in forging a meaningful and productive career, for me, in natural sciences.

**Suzan Mazur**: When will you complete your PhD, and will you stay in the US or return to India following your studies?

**Ramray Bhat**: I expect to complete my PhD within the next two years. I haven't really thought about a next step, as I want to concentrate on the present, which is actually a fantastic period of my scientific life. However, I feel strongly about imbibing all that I can from this scientific environment and country and returning to India where I can contribute in a meaningful way to Indian science, research and education.

# Appendix D:

# Piattelli-Palmarini: Ostracism without Natural Selection

*May 9, 2008*
*12:20 pm NZ*

**Suzan Mazur**: In the book you're writing with philosopher Jerry Fodor on evolution without adaptation, do you share his view that we need a new theory of evolution and that the theory of natural selection is "wrong in a way that can't be fixed"?

**Massimo Piattelli-Palmarini**: Yes, I do. Of course, there is natural selection all around us (just think of the flu virus, mutating and adapting every year, to our detriment) and inside us (just think of our antibodies and our synapses and the pancreas cells and the epithelial cells). The point is, however, that organisms can be modified and refined by natural selection, but that is not the way new species and new classes and new phyla originated.

For that, major changes in regulatory genes and in gene regulatory networks have to occur. All this is perfectly naturalistic and now well documented. Minor changes in the order of activation of master genes can create vast discontinuous morphogenetic changes. Very similar (in the jargon orthologous) genes in insects and in vertebrates produce an inversion in the development of the nervous system.

In essence, in insects the system is ventral, in vertebrates it's dorsal. Two opposite gradients of morphogenetic factors (one the mirror image of the other) produce this difference. Huge difference to the eye, but minor in its origin early in the development of the embryo.

There will be one day, decades from now, I am persuaded, hanging on the walls of the schools, some equivalent of the Periodic System (Mendeleev Table) showing how these genetic regulatory switches combine to give the different forms of life (this is just a metaphor, of course, but I bet it will become a detailed plan one day).

**Suzan Mazur**: Jerry Fodor told me you were handling the biology for the book, but you also have a PhD in physics in addition to being a cognitive scientist. Do you have a hypothesis on origin of form? And would you tell me a little about what you're teaching your class there at the University of Arizona on form?

**Massimo Piattelli-Palmarini**: Yes, I have a doctorate in physics, quite rusty now. And yes, I think that there are fundamental physical and chemical principles operating inside all living systems and partially responsible for the forms of living organisms. Only partially, of course, but at a very fundamental level.

More and more papers, from different quarters (laboratories and researchers still that remain for the most part isolated from one another), show that there are physical principles of optimization, and of optimal compromise, acting on biological forms.

For instance, Christopher Cherniak and colleagues at the University of Maryland have computed literally millions of alternatives to the way the nervous system is organized, from the ganglia of the earthworm (the nematode) to the auditory cortex of macaques, and found that none of these can improve to what we have in reality. Nature has found an optimal solution for the density of connections that is better than the most advanced engineered chips we find on the market today.

A few years ago, in Santa Fe (yes, the Insitute so dear to you and Kauffman), West, Brown and Enquist discovered that the natural ramifications in all circulatory systems (the sap and lymph circulation in trees; the veins, arteries and capillaries in mammals) follow a maximal fractal law. Best transport with minimal distance. Something that evolution has "rediscovered" over and over.

Other instances of optimization are found in other components of biological systems. In phyllotaxis (the geometry of leaves and of flower petals), we see reproduced the Fibonacci perfect spiral, our phalanges have lengths of 1, 2, 3, 5, 8 (the Fibonacci series), and so on.

Now, it just cannot be the result of natural selection that biological forms show the same forms we also witness in spiraling minerals and in spiral galaxies. And when we find a "solution" in living being that turns out to be optimal with respect to many millions of conceivable (and computable, these days, with fast computers) alternatives, it cannot have been selected out of random trials. There have not been dozens of millions of generations of macaques trying out all sorts of cortical patterns of connections, such that only the best survived. That's ridiculous.

**Suzan Mazur**: Why has American science been slow to accept a reduced role for natural selection in evolution? Is it the physics that people just can't grasp?

**Massimo Piattelli-Palmarini**: It's not just American science, but rather Western science, though indeed France has, in this respect, a different story, not quite a noble one.

Some consider Darwinism to be quintessentially "Britannique" and they had Bergson suggesting a different approach to evolution, then the mathematician Rene' Thom and his school, stressing the role of topological deep invariants. They may have come to anti-Darwinian conclusions for rather idiosyncratic reasons.

Anyway, even if we take the many, many biologists in many countries who have contributed to the new rich panorama we have today of non-selectionist biological mechanisms (including the masters of the Evo-Devo revolution), they are reluctant, in my opinion, to steer away from natural selection. They declare that the non-selectionist mechanism they have discovered (and there are many, and very basic) essentially leave the neo-Darwinian paradigm only modified, not subverted.

I think that abandoning Darwinism (or explicitly relegating it where it belongs, in the refinement and tuning of existing forms) sounds anti-scientific. They fear that the tenants of intelligent design and the creationists (people I hate as much as they do) will rejoice and quote them as being on their side. They really fear that, so they are prudent, some in good faith, some for calculated fear of being cast out of the scientific community.

There are, however, also biologists who do not fear to declare, as Gregory C. Gibson (the William Neal Reynolds Distinguished Professor of Genetics, North Carolina State University) wrote in *Science* (2005), reviewing a book on robustness and evolvability: "[this book] contributes significantly to the emerging view that natural selection is just one, and maybe not even the most fundamental, source of biological order".

"Robustness must involve non-additive genetic interactions, but quantitative geneticists have for the better part of a century generally accepted that it is only the additive component of genetic variation that responds to selection. Consequently, we are faced with the observation that biological systems are pervasively robust but find it hard to explain exactly how they evolve to be that way". G.C. Gibson SYSTEMS BIOLOGY: The Origins of Stability. *Science,* 310 (5746), p.237.

And the distinguished evolutionary biologist Massimo Pigliucci, in an excellent book co-authored with the philosopher Jonathan Kaplan, writes:

"It is unwarranted to think that adaptation, diversification and evolution more generally are closely related phenomena that take place via the same mechanisms in the same populations [. . .] Adaptation can, and verifiably does, take place without speciation, as does nonadaptive evolution more generally". Massimo Pigliucci & Jonathan Kaplan (2006), *Making Sense of Evolution: The Conceptual Foundations of Evolutionary Biology.* Chicago: The University of Chicago Press. (Chapter 9, Box 9.2).

There are other expressions of discontent with canonical neo-Darwinism, but, all in all, prudence prevails.

**Suzan Mazur**: It's interesting that there's been a meeting of minds among biologists, philosophers and linguists about language in evolution. Didn't you, MIT linguist Noam Chomsky and the late paleontologist Steve Gould at one point all share similar thinking that language was due to laws of structure and growth and not natural selection?

**Massimo Piattelli-Palmarini**: Yes, that's exact. Gould's untimely death did not allow him to develop fully this side of the issue, but cogent reasons for not trying to explain cognition and language along neo-Darwinian lines have been iterated by his main co-author, Richard Lewontin.

Steve and I taught twice, several years ago, at Harvard, a course together and we had mighty opponents attending it (notably Steve Pinker and Daniel Dennett) giving life to quite animated discussions.

Chomsky gave a guest lecture in that course, with the Nobelist David Hubel also attending (you see how lucky I have been, a dwarf among such giants) and the debate was intense, though friendly. In essence, at the time Chomsky cogently argued (and I reinforced this in some publications of mine) that the very structure of language has peculiarities that cannot have been shaped by (naturally selected as) sub-products of communication, not thinking. They appeared then (around 1985) too idiosyncratic to be the result of a functional selection.

Today (ever since, approximately 1995) the message is basically the same, but with a change of emphasis. There appear to be in language, at a suitably high level of abstraction, properties of elegance and maximization that explain those peculiarities as applications. What once were considered to be explanations (the principles of a restricted set of syntactic modules) now are considered themselves data to be explained. This is called the Minimalist Program, something that has fascinated many linguists, who have engaged with Chomsky (though sometimes there are points of technical dissent) in this program very thoroughly, with what I consider to be very interesting new results.

But it has distanced, in some case even repelled, other linguists. Time will tell whether this research program is right or wrong. I think it's right, though still very preliminary. At any rate, there is no place in this program for any adaptationist, gradualist, neo-Darwinian explanation. This much has not changed.

**Suzan Mazur**: Do you also think that structure, *i.e.*, form is due to a language as Stuart Newman and Ramray Bhat hypothesize in their recent *Physical Biology* paper "Dynamical patterning modules: physico-genetic determinants of morphological development and evolution?

Newman & Bhat describe the role of a pattern language – DPMs (dynamical patterning modules) – in the self-organizing of all 35 animal phyla by the time of the Cambrian explosion a half billion years ago.

Would you comment briefly on the Newman & Bhat paper?

**Massimo Piattelli-Palmarini**: I find this kind of morpho-dynamic approach immensely productive. The term "language" could be left out, but the basic ideas are right and very interesting.

Similar ideas have been expressed, among others, by Eric H. Davidson and collaborators (at Caltech) arguably the leading expert today of genetic regulatory networks.

Modularity and entrenchment in development are mature fields and discoveries towards discontinuous patterned changes in developmental constraints are being published very month (see Psujek, S. And R. D. Beer (2008). "Developmental biology in evolution: evolutionary accessibility of phenotypes in a model evo-devo system." *Evolution & Development* 10 (3): 375-390. Just published).

A biochemist at Boston University, Michael Sherman, has proposed the idea of a "universal genome", so akin to Chomsky's idea of a universal grammar (then unbeknownst to Sherman) that Michael is now reading some minimalist linguistics and tells me he finds that field extremely interesting and congenial.

Symmetrically, Chomsky told me about Sherman's approach that he thinks decades from now will become "mainstream biology" (his words).

**[Noam Chomsky emailed me that he was not endorsing the Michael Sherman paper. He noted that Sherman's ideas were "plausible" but that "serious commentary" on specific**

**proposals requires more background and knowledge than I have".**

**He also said: "Since the '50s, I've assumed (and occasionally written) that something like the Thompson-Turing approach, with its roots in rational morphology, ought to be the right track…".**

**Curiously, however, it is Stuart Newman's and Ramray Bhat's theory of form paper based on DPMs (Dynamical Patterning Modules) that's in the D'Arcy Thompson – Alan Turing tradition, not Michael Sherman's on the Universal Genome. –SM]**

As I said earlier, my way of depicting what will happen decades from now, is a sort of universal chart of morphogenetic pathways that will be displayed like today's Mendeleev's table is. I think too that this will be the mainstream biology of the future.

Look, when Sherman stresses that the sea urchin has, in-expressed, the genes for the eyes and for antibodies (genes that are well known and fully active in later species), how can we not agree with him that canonical neo-Darwinism cannot begin to explain such facts?

**Suzan Mazur**: Who are some other evolutionary thinkers with views along this line you find interesting?

**Massimo Piattelli-Palmarini**: Well, of course, Stuart Kauffman, whom you have recently interviewed, a pioneer in the search for physical bases of biological morphogenesis (his name and his earlier work were pointed out to me by Steve Gould around 1985). Eric Davidson, mentioned above, Gerd Müller and Stuart Newman, just mentioned, Gunter P. Wagner and the whole idea of entrenchment and modularity, Massimo Pigliucci, and of course the main authors in Evo-Devo (for instance Sean Carroll, Mary Jane West-Eberhard and

Marc W. Kirschner), though they are sometimes over-prudent in keeping within a neo-Darwinian frame.

Not many of them, with the exception of Kauffman, point towards physical invariants in morphogenesis, but their important data offer the basis that any such approach will have to take into consideration.

**Suzan Mazur**: Do you consider self-organization or autoevolution, as cytogeneticist Antonio Lima-de-Faria calls it, a kind of self-determination? And if so, why would people resist that idea regarding evolution?

**Massimo Piattelli-Palmarini**: Well, Lima-de-Faria wants to do without selection altogether, an extreme view.

The difficult issue, as Kauffman had emphasized years ago, is to integrate physical principles, genetics, development and different kinds of selection, acting in different ways at different levels.

Self-organization is of course an important component, but not much has been discovered beyond generalities. The immense amount of intricate detail that geneticists and developmentalists have been discovering over the years dwarfs general metaphors like autoevolution and even self-organization.

The challenge now is to integrate, not to substitute these metaphors for hard work over many years.

**Suzan Mazur**: Do you think the Konrad Lorenz Institute's July symposium about an Extended Evolutionary Synthesis – where theory of form and non-centrality of the gene will be an important part of the discussion – will help to steer the scientific community and the public toward a better understanding of how form arose without selection?

**[Stuart Newman will present his paper, co-authored by Ramray Bhat, at the KLI symposium in Altenberg on how all**

**35 animal phyla self-organized a half billion years ago by the time of the Cambrian explosion, with selection coming into play as a "stabilizer" after the highly plastic multicellular organisms formed.]**

**Massimo Piattelli-Palmarini**: We said [evolution] without adaptation, not without selection.

There is selection, there has to be selection, though not the macroscopic, uni-level selection of classical neo-Darwinism. How genes interact with the physical factors we saw earlier here still remains to be determined. How evolution and selections (plural here) "ride" so to speak the narrow channels of what is physically possible is still a mystery.

It does not help to depict the genes as inert stuff, dead material. Of course the whole cell is needed for them to be activated, expressed and so forth. But genes can be transplanted, cut, spliced, duplicated, etc.

It's silly to preach anti-geneticism. The real new synthesis will have to be between all these components, none excluding the others.

**Suzan Mazur**: How long do you estimate it will take for theory of form to be understood and gain credence within the scientific mainstream?

**Massimo Piattelli-Palmarini**: Well some 20 years for the elite of the scientific community. Maybe 50 before it becomes high-school textbook material.

**Suzan Mazur**: Developmental biologist Stuart Kauffman, one of the pioneers of self-organization, rejects reductionism in his new book, *Reinventing the Sacred,* saying that a couple in love walking along the Seine are not just particles in motion. What are your thoughts about this?

**Massimo Piattelli-Palmarini**: I like him and his work, but this is not a sensible remark. Who ever claimed they are? Perfusing his quest for the basic laws of morphogenesis with this kind of holistic humanism does not help, sorry.

More interesting is to ask whether a chromosome is a giant molecule or a society of interacting modules. We do not know zilch about the meaning of the number of chromosomes in the different species.

That number cannot be altered (chromosomal aberrations in humans, even minor ones, produce well known syndromes, some lethal), but nobody has an explanation of what it means for us to have 46, for the platypus to have 52 (of which 10 are sexual), and for the chimp to have 48.

One day, I bet, these data will be part of the mural table I anticipate, but as of now no one has the faintest idea.

Reductionism is bad when it's bad, when it destroys what is to be explained. But it's mighty good when it's good, when the assembly of the parts does explain the property of the whole.

I resist humanistic anti-reductionism. Without intelligent reductionism we would not have the science we know and like. Mendeleev's table is also a kind of reductionism, a welcome one.

**Suzan Mazur**: Of the theories on origin of form out there at the moment that you've reviewed – which do you find most plausible?

**Massimo Piattelli-Palmarini**: Factorial discrete changes in the regulation of master genes, and topological invariants in living forms. A bit of what D'Arcy Thompson and others had in mind, but with close integration of genetic regulatory networks. No simple overarching solution will work. Many factors will have to be integrated.

**Suzan Mazur**: Have you seen any convincing new illustrations and evolutionary models?

**Massimo Piattelli-Palmarini**: The idea of a universal genome, a' la Sherman, is the most interesting I have seen recently. Not a single key, but an important key.

**Suzan Mazur**: Would you comment on Stuart Pivar's animations of body parts?

**Massimo Piattelli-Palmarini**: Very interesting, though his idea that the torus is the mother of all forms is not persuasive. There must be a dozen of such mothers, not just one.

**Suzan Mazur**: When do you expect your book will be published?

**Massimo Piattelli-Palmarini**: Sometime in late 2009. But mind you, it contains some of these ideas, but also other important ideas I did not mention here. Notably that the strict analogy between Behaviorism and neo-Darwinism is quite fatal to the second, though few seem to have noticed.

How can the first be agreed to be defunct but not the second? Also, the ineliminable intensional (mentalistic in some unrecognized way) character of notions such as selected for and ecological niche. But for these, you have to wait to read it.

Finally, it will be called *What Darwin Got Wrong*. Not final yet, but probable.

## Appendix E:

# Niles Eldredge, Paleontologist

*February 13, 2008*

Phone Conversation

**Suzan Mazur**: So you don't think there's any alternative to natural selection? You call them "additonal ways".

**Niles Eldredge**: I don't think there's any alternative. You know, I'd have to. We hung up. I said to myself, I don't even know much of the literature of self-organization at all, so I can't even really speak about that.

**Suzan Mazur**: Body form arises from a pattern in the cell membrane not from DNA.

**Niles Eldredge**: Right. But how do you modify that through time and so forth. Yeah, I'm sure that there's something to that. It's actually nothing that I'm really familiar enough with to even be quoted on.

**Suzan Mazur**: Well, what about the... I was talking with Massimo Pigliucci. Do you know him? He's a geneticist out at Stony Brook.

**Niles Eldredge**: Yeah. I know of him.

**Suzan Mazur**: He was saying he sticks with natural selection but thinks we've gone as far as we can go regarding finding any more genes for humans. And we need to explore other areas. He thinks self-organization is really worth looking at.

I was also talking with philosopher and zoologist Stan Salthe. Do you know him?

**Niles Eldredge**: Yes.

**Suzan Mazur**: He describes self-organization as "up & coming".

**Niles Eldredge**: You might want to talk to Dan Brooks at the University of Toronto about this.

**Suzan Mazur**: Dan Brooks.

**Niles Eldredge**: Yeah. He's much more up on this stuff than I am.

**Suzan Mazur**: Right. But you worked with Steve Gould on punctuated equilibrium. You wouldn't characterize Steve Gould as being part of the self-organization school?

**Niles Eldredge**: I certainly wouldn't and he would have said to you what I said. If he were alive right now, he'd probably be more up to date on that stuff. Because these things intrigued him. But, if you want to see what I say to Steve and about punctuated equilibrium, you can check it online at – I have an online article on – we have a new journal called *Evolution, Education and Outreach*. We're trying to connect the lab with teachers.

I'm editing this journal with my son, who's a teacher. And if you go to *Springer.com*, you'll find that journal. The next issue is partially up. It would be Volume I, no. 2. And you'll find – I just looked at it myself and downloaded a copy – my article about the early evolution of punctuated equilibrium.

You're going to find out that it's very conventional. It's talking about speciation. It's talking about natural selection. And it's talking about stasis, which is the tendency of species not to change that much once they first appear.

**Suzan Mazur**: I'll definitely take a look at it.

**Niles Eldredge**: But it's not the radical sort of thing. I mean it was radical in the sense that everyone was a gradualist. But there was a lot of smoke that was generated into the fire that was there. When our essay came out in 1972.

Apparently the blogosphere in the last couple of weeks has been full of this all over again, that punctuated equilibrium was a macromutational theory and all this and that and it's simply not true.

**Suzan Mazur**: What about Richard Lewontin's story in the *New York Review of Books* about Steve Gould?

**Niles Eldredge**: Somebody sent that to me... Steve was interested in exploring new ideas. He was not saying things for the sake of saying them, which is almost what people are accusing him of having done. And the thing that he was best known for was that paper that I did with him, which was actually based on an earlier paper that I wrote. And I'm telling you – it's very conventional biology. It just seemed unconventional at the time.

And then Steve also had written separately in *Natural History* magazine seven years later or so about Richard Goldschmidt's theory of macromutations. So he suggested in there and I think he was one of the earliest ones to do this. This is really like evo-devo, not self-organization.

He suggested that if you get slight changes in the regulatory apparatus of the genome, which was then only beginning to be known, that it might have cascading relatively larger effects in the phenotype. And so that was macromutation. And that's what the blogging is about because people say punctuated equilibrium was a theory of macromutation and it certainly was not. It was a theory of speciation for natural selection.

**Suzan Mazur**: So you don't find it problematic that we've only found 25,000 genes or so and that we're not going to find anymore? That we have to look in another direction for answers?

**Niles Eldredge**: Well, I'm really not a geneticist so I really don't know. But yes I have to assume that's largely correct. Are there cytoplasmic effects of development? Sure. How they get ensconced so that all the members of a single new species all look alike and it's different from their ancestor if it's not genetic, I'm not sure.

**Suzan Mazur**: You don't believe natural selection had no role in form?

**Niles Eldredge**: Absolutely not. I certainly don't believe that it had no effect. I'm a very conventional evolutionary biologist. I disappoint people sometimes.

**Suzan Mazur**: In January, the National Academy of Sciences came out with *Science, Evolution and Creationsim.* It's a very general book. The people behind the discoveries are not really discussed to any extent. The books of Steve Gould are listed, but not his controversial writings. Do you find any problem with the approach of that book?

**Niles Eldredge**: I've seen it. I haven't studied it. I gave it to somebody to write a review for our journal.

**Suzan Mazur**: So you haven't heard any buzz about its "simplicity".

**Niles Eldredge**: No. I mean, look, when you're fighting school boards who want to adopt Intelligent Design, you've got to write in very basic terms. It is a political problem. And there's always a problem, as you know… in communicating science to the public and being clear about it.

**Suzan Mazur**: But the thing is they didn't put in any, as you termed them, "additional ways" to consider regarding evolution.

**Niles Eldredge**: No, because it's all regarded as speculative and on the forefront and stuff… What they're trying to do is say where we are now, where we're comfortable, where we can actually say that this is the way people really do think for the most part.

**Suzan Mazur**: Meantime some of the public money that has gone to NAS wound up in secret arms research, which is something Richard Lewontin resigned from NAS over.

**Niles Eldredge**: You have to ask him about that. Yes, I think he did. So what does that have to do with...

**Suzan Mazur**: No, these are issues.

**Niles Eldredge**: I think the problem of people trying to teach religiously-based ideas as if they were true science and valid science in the classroom is a serious issue in and of itself. And I'm glad that the National Academy of Sciences once again tried to do something to provide some material so people can reach an *informed decision* in school boards around the country. How effective it is, I couldn't tell you.

**Suzan Mazur**: On the other hand, the *Astrobiology Primer* that NASA/NAI supported in 2006 and is supposed to be updated next year – the editor-in-chief is an Episcopal priest.

**Niles Eldredge**: Uh huh. So.

**Suzan Mazur**: At the time he was a seminary student. I find that kind of unusual. Inside there's no mention of alternative or additional ways of evolution. It's about natural selection. It does mention Komura [neutral selection]. But there's no real space given to "additional ways". And there's no mention of you and Steve Gould.

**Niles Eldredge**: So they're giving the conventional view. If you open Doug Futuyma's book, the guy at Stony Brook who is probably one of the most famous evolutionary biologists in the country now, if for no other reason than he wrote that widely used textbook – you're not going to find that Steve and I get a very good shake in that book. And you're not going to find, I don't think, an extended discussion of self-organization, if it's even mentioned. So this is not government. This is Doug Futuyma selling textbooks to the kids of the United States…

**Suzan Mazur**: Do you see it as problematic that scientists from one field, say in biology, and scientists from mechanical engineering, chemistry, *etc.*, are having a difficult time communicating?

**Niles Eldredge**: Absolutely. It's a problem. I'm a paleontologist and most of the people who make a living as evolutionary biologists, teaching it or whatever, are geneticists. There's a big problem because we don't speak each other's language all that fluently.

It's a struggle. That's why I've been reluctant to comment on some of the things you're bringing up because I'll tell you right off – I don't know that much about it.

**Suzan Mazur**: Because there might be something happening there but not everyone feels comfortable talking about it.

**Niles Eldredge**: Exactly.

**Suzan Mazur**: Maybe we need to move in the direction of a new language.

**Niles Eldredge**: Absolutely. I wrote a paper for our new journal called "Hierarchies and the Sloshing Bucket, Toward the Unification of Evolutionary Biology". In it I talk about how difficult it is for us to talk to one another because we're dealing with phenomena that are these diverse spatio-temporal scales. Everything from molecules to ecosystems. Species to phyla. We're talking about all of geological time.

Some people are only looking at the present. They're not trained to look at the past. Other people are inferring the past by looking at the present. Other people are looking at the past, the fossil record.

And these are fields that are based on real phenomena, real systems that exist at these different spatio-temporal scale. So the first thing we have to do in trying to communicate, to come up with a more complete evolutionary theory, is to realize that there are all of these kinds of phenomena, these entities and systems and the interactions between them. You have to acknowledge that first, and then you have to decide – well how are you going to go about…

I don't think anybody says Darwin actually grasped what was known in his own lifetime about biological systems in their entirety. And now we know more – about microbiology and molecular. There was nothing known about genetics in Darwin's day at all… I've never met anybody who could grasp all this stuff.

**Suzan Mazur**: Is it possible to communicate about the evolution debate to the general public in a significant way? Some think it can't be done.

**Niles Eldredge**: Yes. That's why founded this journal. I grew up in an era when people hardly paid any attention to communication at all. And it was left to people like Isaac Asimov and some really good writers to do it. And they either succeeded or failed according to their own skills and the luck of the marketplace.

And now you cannot get a grant from the federal government in what I would call evolutionary biology – I think this is true – from the National Science Foundation – you cannot apply for funding for research in evolutionary biology now without having an outreach component to your proposal.

**Suzan Mazur**: What do you mean by an outreach component?

**Niles Eldredge**: You've got to explain in the body of the proposal how you're going to educate other people beyond the confines of your narrow discipline.

**Suzan Mazur**: Also, everything is getting so visual. Are you familiar with Stuart Pivar's torus model? ...

**Niles Eldredge**: You're talking about morphology and its modifications. If you look at D'Arcy Thompson's stuff from the early part of the 20th century, you'll see the same thing. He was deforming fishy-looking fish into ocean sunfish by just deforming the coordinates, but that doesn't really tell you how it happened. It's a way. It's a very clever way of describing what did happen, but it doesn't tell you how it happened.

When Darwin came along, most of the anatomists aside from Huxley didn't even go for it because they thought that basically the anatomical configuration of something like a human being or a leopard, or something like that, was so built into the program they had no idea what the program was. But it would be hard to modify any of the parts without having the whole thing messed up.

So anatomists are very conservative. But then some of them got bitten by the bug. They understood that evolution must have happened and they still tend to be sort of describing static states and how they get deformed and modified into other ones. But they are not theories of how that happened. Or not convincing ones.

**Suzan Mazur**: This experiment back in the 50s – Miller-Urey where they created amino acides in the laboratory. Could you address that in relation to the current evolution debate?

**Niles Eldredge**: They made chains of amino acids. Not even all functional proteins. But I do think that that sheds a lot of light on the early chemistry of the Earth. They did not create life however.

Now there are other ideas about the importance of clay molecules and shaping DNA. And a big argument is whether RNA came first. Proteins afterwards or vice versa and so on and so forth. And it's not nailed down.

The other thing I'd like to say about the origin of life, because you just shifted the frame of reference here, and I think Darwin was right when he said that the origin of life is a different subject from the consequent evolution of life. That's a molecular biology, biochemical problem basically. The origin of the early molecular thing that went on to become living systems.

**Suzan Mazur**: Do you know Bob Hazen, the George Mason University mineralogist? He seems to be open to the idea of self-organization.

**Niles Eldredge**: Yes. I know Bob. I like him. Right. He's a mineralogist. He's not an evolutionary biologist. So be careful.

He's been in my house here where I'm sitting now. I'm looking at a huge collection of antique cornets that I own. And I sold one to Bob. He's a great trumpet player also. But he's not an evolutionary biologist.

# Appendix F

# Stan Salthe: Neo-Darwinians Risking 'Rigor Mortis'

"The question is: Does the neo-Darwinian program in population genetics, oriented around the concept of natural selection, serve the broader interests of diverse evolutionary biologists such as naturalists, paleontologists, systematists and Evo-Devo workers? Put otherwise, is that discourse still worthy of the central place it has had in evolutionary biology?

I would suggest that its work has essentially been accomplished, and that it is now becoming a highly esoteric mathematical discourse about less and less, working its way onto an asymptote of what it is capable of contributing. Nevertheless, it has left at least one problem hanging -- the simultaneous evolution by selection of traits involved in several to many independent biological functions. (I do not refer here to selection entraining simultaneous changes in multiple genes, which is well known and exemplified in the programming trick, 'genetic algorithms', which works many genes on one function.)

This problem was first raised some time ago by J.B.S. Haldane, as the 'cost of natural selection', and it was 'solved' by several evolutionary biologists – viz the 'soft selection' of Bruce

Wallace and the 'threshold selection' of John Maynard Smith among several others.

This general solution, however, as a holistic one, forfeits discussing independent traits altogether. And so, statements like 'evolution of the eye' would not be supported by population genetics unless visualized as the sole evolving function in very large populations. It may be that such concepts as the evolution of the eye are misleading, but this is generally not clearly cautioned against in evolutionary texts."

Another outstanding problem not dealt with in the evolutionary biology dominated by neo-Darwinians is convergent evolution and ecological equivalence. The Darwinian program aims to discuss 'descent with modification' and addresses the evolution of diverse and disparate types. It is not suited, unaided, to handle convergence or the evolution of increasing complexity, both of which are macroevolutionary problems.

Population genetics deals with microevolution – the change in gene frequencies from one generation to the next. Even there, some of its versions (*e.g.*, the Dobzhansky-Lewontin school) concern, not evolution *per se,* but the problem of preventing population extinction.

By being 'rigorous' the neo-Darwinians have convinced other scientists of their validity as the foundation discourse of evolution studies, but their 'rigor' pertains to issues very far from the interests of most evolutionary biologists, risking 'mortis'."

Stan Salthe, the philosopher and zoologist who brought us the Fodor email chain (Chapter 3 –"Jerry Fodor and Stan Salthe Open the Evo Box"), told me you can't dismiss the censorship going on in the evolution debate.

He sent me his correspondence with the neo-Darwinian journal *TREE (Trends in Ecology and Evolution)* in which he asked *TREE* to publish his letter arguing to "save the phenomenon of convergent evolution even if it seems inconvenient" to the Darwinian perspective on organic evolution. Salthe was responding to an article *TREE* published suggesting the concept of convergent evolution be eliminated based on a totally genetic analysis. *TREE* refused to publish Salthe's letter, however.

Salthe says convergent evolution happens when different species become similar without involving "similar genetic representation". Examples of this, he notes, are old world vultures evolving from hawks and new world vultures evolving from storks. Here's Salthe's letter to *TREE* and its responses:

*10 February 2008*

> Dr. Katja Bargum
>
> Acting Editor
>
> Trends in Ecology & Evolution
>
> Sir/Madam
>
> In Volume 23, Number 1, there is an opinion paper by Arendt and Reznick that I think requires comment, in that it redefines Convergent Evolution out of existence. Would it be appropriate/acceptable to submit a short rejoinder of just a few paragraphs?
>
> Stan Salthe

Biological Sciences

Binghamton University

*11 Feb. 2008*

Dear Dr. Salthe,

Many thanks for your email. In the case of responses to published articles, we generally encourage authors to submit the letter they would be considering, or a summary of their argument. If you would like to send us such an outline, I would be glad to consider it further for publication.

Best regards

Katja

Salthe sent in his outline. "Concerning A Proposed Deconstruction of Convergent Evolution" – S.N. Salthe, February, 2008

**Concerning A Proposed Deconstruction Of Convergent Evolution.**

S.N. Salthe, February, 2008.

Arendt and Reznick (*TREE* 23:26-32) have presented an argument for eliminating the label 'parallel evolution', which involves changing the terms of the comparison between convergent and parallel evolutions from the phenotype to the genotype. While parallel evolution has mostly been viewed as a largely genotype focused concept, convergent evolution has not been formulated that way (see below). The authors manage to make their argument seem plausible by moving the discussion from the phenotypic scale, where parallelism and convergence have traditionally been described, to that of the genetic material, focusing upon individual loci. At that level, they find that the distinction between parallel and convergent evolution effectively disappears, since the effects of the regulation of related single genes can be similar even in distantly related organisms.

On genetic explanations, it has occurred to me that such convergent similarities as that between, *e.g.*, the focusing eyes of cephalopods and vertebrates might some day be assimilated to genetic discourse by way of the lateral gene transfer concept. Interestingly, however, in this case the lens crystallins are known to be made from unrelated proteins, and so by the authors' criteria there would be no 'convergence' here, even though in the classical definition (see below) this difference would be a crucial fact of convergence, almost diagnostic of the concept.

The authors' suggestion is in line with the long drift of modern evolutionary biology toward exclusively genetic explanations, but the phenotype is now being brought back into focus in evolution discourse by Evo-Devo, and, ironically, by way of Dawkins' 'The Extended Phenotype", as viewed through the works such as those of Scott Turner.

The authors argue that parallel and convergent evolutions can be subsumed under one label, 'convergent evolution' - a curious choice given the typical scope of these concepts until now.

Classical definitions:

Parallel evolution: similar evolutionary changes among relatively closely related linages resulting from similar selection pressures entraining similar tendencies toward developmental explorations, based on closely similar genetic information. We could describe it as 'becoming similar together'.

Convergent evolution: similarities among quite to very distantly related clades, which is unlikely to involve similar genetic representation (as in the lens crystallin example). We could say 'becoming similar completely independently'.

We may note again that traditionally convergence was clearly not a genetic condition.

My point here is that the authors' suggestion works to eliminate an important distinction, and to obscure an interesting reality that ought to be retained as potentially soluble in a future evolutionary biology. Here are few examples of convergence at different scales:

> the focusing eyes of cnidarians, cephalopods and vertebrates
>
> old man's beard (lichen) / Spanish moss (bromeliad)
>
> hagfish / an Amazonian deep water catfish
>
> Polyodon / the whale shark
>
> Old World vultures (from hawks) / New World vultures (from storks)
>
> weasels / civets

cacti / euphorbias

hummingbirds and hummingbird moths

nests with chimneys in both ants and termites

some metatherian lifeways / some eutherian lifeways

The same body forms of: paleoniscoid fishes / holostean fishes / teleostean fishes (iterative evolution)

pine barrens vegetation in northern and southern eastern North America

Mediterranean vegetation in the Andean Puna and in Australia

Thus, convergent effects reach all the way to the biome level – *e.g.*, the Mediterranean vegetation communities. While selection acts on genes, vegetation similarities in communities are clearly supra genetic.

Similar phenotypic changes of a very general kind, like body shape streamlining in many distantly related aquatic species, has not been under consideration for convergence since it is easily conceived to be mediated by selection in any kind of organism that must swim swiftly in water. This constraint eliminates the otherwise interesting cases of ichthyosaurs and porpoises and of fishes and squids. The kinds of phenomena suggesting a need for new concepts can be illustrated by a comparison of seahorses with chameleons:

camouflaged bushwhacking mode of stealthy hunting

very rapid strike at prey

slow progression, using prehensile tail to grasp vegetation

independently moving eyes

live birth (only in some chameleons)

The extreme difference in habitat as well as the significant difference in taxonomic distance in this case magnifies the import of these similarities.

I would argue that we ought to 'save the phenomenon' of convergent evolution even if it seems inconvenient to the dominant Darwinian and cladistic perspectives on organic evolution. The fact that convergence is not in conformity with the 'descent with modification' model of either the neoDarwinian or cladistic programs signals rather that it is an outstanding problem in evolutionary biology. Given the divergence model of organic evolution, there should be no reason to suppose that selection would arrive at similar solutions to similar selective pressures except for very general solutions like streamlining in fluids or in very closely related forms. I think that the major significance of the Arendt and Reznick paper is that single loci can be regulated in ways independent of our classifications of the organisms carrying them.

*TREE* responded:

*Tuesday, March 4, 2008*

> Dear Stanley,
>
> Many thanks for providing an outline for your letter. However, I regret to say that we have already commissioned a letter presenting a very similar counter-argument to Arendt and Reznick, and this precludes one from commissioning another article on this point.
>
> I am sorry to disappoint you. I intend no criticism of your work, and wish you the best of luck in publishing your letter elsewhere.
>
> Very best wishes,
>
> Katja

Salthe thinks the *TREE* people went to his web site and saw his critique of natural selection and reacted. "And that is it!" – he said. "You know, forget it! This guy is poison." He then wrote back to *TREE* saying: "I am glad you have found an appropriate author for this idea."